EINSTEIN'S TABLE:

The Search to Find a Cure for Chronic Hepatitis B

KIMBERLY D. JUNGKIND, MPH, RN, BSN

ISBN: 1468022075
ISBN-13: 9781468022070
Library of Congress Control Number: 2011962155
CreateSpace, North Charleston, SC

DEDICATION

This book is dedicated to Baruch S. Blumberg, MD D.Phil., who passed away during the writing of this book. He provided passionate support for the Hepatitis B Foundation since its inception. He offered guidance, ideas for this book through several hours of recorded interviews and ongoing phone conversations. His efforts to assist the Hepatitis B Foundation and share his vision will be remembered by many. Appendix 1. of this book contains a tribute to Dr. Blumberg from four of his closest colleagues, including Timothy Block, PhD, W. Thomas London, MD, Professor Raymond Dwek, FRS, and Hie-Won Yvonne Hann, MD.

> *"The thrill of discovery and science for its own sake*
> *are attractive parts of the scientific endeavor,*
> *but we are now*
> *ready to help apply our discoveries*
> *to matters of life or death."*
> **Baruch S. Blumberg, MD, DPhil.**

This book is also dedicated to my husband, Donald, and my children, Daniel and Rachel, for their support and understanding as many hours were spent in the home office poring over information and adding content to the book over many months. I appreciate their love and patience with me as I tried to balance the whirlwind of work and family activities.

An inspiring story of a small group of concerned citizens committed to finding a cure for chronic hepatitis B, a disease that affects one in twenty Americans. Their humble beginnings and remarkable efforts over the past twenty years have expanded education, improved outreach, developed resources, engaged support, drafted legislation, created an award winning website and brought together the best researchers to find a cure for chronic hepatitis B—all to make a difference for patients nationally and internationally through the achievements of the Hepatitis B Foundation.

On the previous page was an image of a slanted "B" with the spinning sphere at the bottom. The image was shared in a correspondence from Dr. Blumberg to Dr. Block years ago. Dr. Blumberg credits Michael Gibson, a former student, for preparing this image for a hepatitis B awareness week in an effort to recommend extending hepatitis vaccination in the U.K.

CONTENTS

FOREWORD

On April 5, 2011, Baruch S. (Barry) Blumberg, Nobel laureate for the discovery of the hepatitis B virus, co-inventor of the first vaccine to protect against infection with the virus, and inventor of the first test for the virus in human blood, passed away suddenly and unexpectedly. Barry was an early supporter of the idea of a Hepatitis B Foundation (HBF) and a mentor for Timothy Block, Ph.D. Barry's goal was to eliminate hepatitis B from the face of the earth, just as had been done with smallpox. The job was harder with hepatitis B because smallpox caused acute, but not chronic infections; therefore vaccination alone could eliminate the disease.

Hepatitis B Virus (HBV), however, causes both acute and chronic infections. In fact, worldwide, 350 to 400 million people are chronically infected with HBV and they are at high risk of developing and dying from cirrhosis or liver cancer. Blumberg realized that two efforts would be needed: 1) a global campaign to vaccinate all newborn babies against HBV to prevent new infections and 2) new drugs that would eradicate HBV in the huge population that was already infected with the virus.

The HBF is dedicated to achieving both goals. It began in 1990 as an idea floated in a conversation between Timothy Block and Jan and Paul Witte. It is now recognized nationally and internationally as the main professional organization for information, advocacy, and research on hepatitis B and the most likely organization to achieve Blumberg's lofty goals. The story of the Foundation and its journey to accomplish these goals is detailed in this book. As stated in its mission, "the HBF is the only national non-profit organization dedicated solely to the global problem of hepatitis B." It is committed to finding a cure for the disease, improving the quality of life of those affected by it, and preventing new infections from occurring.

Through public health and outreach, it promotes disease awareness, supports immunization programs, and is the primary source of reliable information for patients with hepatitis B and their families. It is also the "go-to" resource for the medical community, scientific community and the general public. To find a cure for hepatitis B, it conducts original research in the laboratories of its partner organization, The Institute for Hepatitis and Virus Research (IHVR), focused on discovering both new drugs and new targets for drugs that may not only suppress, but eradicate infections with HBV.

Barry Blumberg was delighted with how rapidly the HBF and IHVR had progressed in achieving the eradication of hepatitis B. With public and private support, the goal of eradication will be achieved in the foreseeable future.

To quote Winston Churchill, "we are not at the beginning of the end, but we are surely at the end of the beginning."

W. Thomas London, M.D.
January 3, 2012

PREFACE

In this book I describe the inspired determination of co-founder Dr. Block in his quest to identify and address the seriousness of chronic hepatitis B and to lead his fellow co-founders on a path toward public awareness, outreach and cure research. In an effort to be as accurate as possible, I have spent hundreds of hours researching over the past year. Much of the information came from books, interviews, news clippings, records, articles, and published literature to accurately capture and convey the contagious compassion behind the story of the Hepatitis B Foundation.

This would not be possible without the assistance of my friends, family, interviewees, contributors, Hepatitis B Foundation board members, founders and employees. They deserve my heartfelt thank you and deep appreciation. I am indebted to them for their encouragement and willingness to share their vast store of knowledge, memories and documentation of Hepatitis B Foundation events and activities. I am particularly indebted to over fifty people whom I spent interviewing and re-interviewing for this book and to the Founders of the Hepatitis B Foundation. They were very patient and allowed me the opportunity to ask many questions, take notes and record the interviews. I have been fortunate to have the support of so many people who have instilled in me the confidence to move forward with writing this book. My hope is that the book will inspire others.

I also want to thank Herman Baron, the president of Diane Publishing and a friend of Barry Blumberg, who has been guiding me in the final stages of the book. He shared his publishing expertise and experience at a key time when it was needed. I appreciate his knowledge, good judgement and willingness to assist.

As a long-standing Board member, I appreciate the approval and support of the Board members to move forward with writing this book. I am grateful for the support of the book from Jean Blumberg, wife of Dr. Blumberg, who has graciously given her time to read the book prior to publishing, provide revisions and her blessing to publish.

INTRODUCTION

Every book holds its surprises and insights along the way when you take the time and reflect. Two insights emerged as I was writing this book: the importance of "it takes a village" style support for the Foundation & its mission and the "Einstein-like" parallels of inspiration.

As I was writing, it became apparent that many people have contributed to the success of the Hepatitis B Foundation. From the Founders, members of the Board of Directors, the Community and Scientific Boards, international and local researchers, office staff, members of the community, advocacy professionals, marketing/communication professionals, political experts, generous donors, friends, family and supporters, etc. –all have been important and key to the success of the Hepatitis B Foundation.

Without all of these supporters, the energy, enthusiasm and advances could not have happened. Many of you have told me how you feel humbled to be a part and contribute to this Foundation that has truly helped so many people across the U.S. and internationally. You have shared with me your challenges to place hepatitis B at the forefront and triumphs you have experienced to move closer to educating the public and finding a cure for chronic hepatitis B.

I appreciate your time and warm welcome into your homes and offices as you let me listen and share your stories through this book. The story of this amazing Foundation's humble beginnings, growth and accomplishments is truly admirable. It is my hope that other organizations will see this Foundation as one to study and learn what was successful—and to inspire! I believe that the Founders enthusiastically encouraged others who, in turn, convinced others to help on behalf of the worthy cause to find a cure for chronic hepatitis B. Inspiration is a key ingredient to the success of any organization or foundation.

"Einstein-like" inspiration was also a part of the start of this Foundation. From a table that later sat inside the guest cottage that Einstein frequented decades earlier, "Einstein-like" parallels emerged as the Foundation grew. Reading about Albert Einstein and reflecting on Dr. Baruch Blumberg's journey related to the discovery of hepatitis B, I found three similar parallels between Albert Einstein, the Founders and Dr. Blumberg. All of these professionals seem undaunted by their journeys leading to their accomplishments.

First, they were all willing to take the time and discover what was needed to solve the difficult scientific challenge that emerged. Albert Einstein was a Nobel laureate who always appeared to have a clear view of the problems of physics and the determination to solve them.[1] Dr. Blumberg, also a Nobel laureate, had a strategy of his own with other close collaborators. He investigated relationships between gene distribution and disease susceptibility which yielded the discovery of the hepatitis B virus and then the vaccine. Timothy Block, PhD., one of the Founders of the Hepatitis B Foundation, was able to visualize the main stages of research and growth that was needed on the way to his goal and he regarded his major achievements as mere stepping-stones for the next advance.

Secondly, Dr. Block was a post doctoral fellow in Biochemistry at Princeton University (where Einstein taught years ago) and was mentored by a Nobel laureate, Dr. Blumberg. Timothy and Joan Block, Paul and Janine Witte – also had a challenge. Their challenge, to cure chronic hepatitis B, required cohesive energy and passion to visualize the next step and move toward achieving that goal. The Founders always moved forward with an unstoppable pace, and continue that pace moving into the future.

Thirdly, Albert Einstein, Dr. Blumberg and the Founders seem to be uncomfortable with personal notoriety and fanfare. They

---

1   Nobel Prize.org http://nobelprize.org/nobel_prizes/physics/laureates/1921/einstein-bio.html accessed: June 28, 2011

would prefer to complete their vision and let the vision take center stage. They were all unafraid to share their vision with those in the political and advocacy arena. In addition, they were all inherently thoughtful and powerful advocates for people and patients in need.

All three of these insights were very interesting to uncover. For the purposes of this book, it was the interplay of these characteristics that would establish the tone and set into motion energy and resources that proves that a group of people really can make a difference. The book contains literary threads of NASA terminology since Dr. Blumberg was the Director of the NASA Astrobiology Institute. It also contains the literary threads of Albert Einstein in the form of a quote at the beginning of each Chapter.

I think if Einstein were living, he may have stopped by the Center or joined in on one of the conversations at the guest cottage in New Hope which he frequented on weekends to play cards, gather insights from people and just enjoy the fun, lively conversations with the Founders and other intellectuals in the community.

Even though Einstein's accomplishments are not detailed in the book, my research, interviews, printed newspaper articles and publications have led me to better understand how this interesting story of the emerging international Hepatitis B Foundation unfolded. Leadership and thoughtful research is important but supportive researchers, friends, relatives and advocates also contribute to the success of the vision. People, who were committed to the vision of the Hepatitis B Foundation, were people who also cared about the well being of other people.

Finally, as a longstanding Board member, I was thrilled to tell the story about this innovative organization and the journey of the leadership, staff, supporters, advocates, community leaders, scientists, researchers and politicians—all who have contributed to it's success. It was a pleasure to bring the stories to life so others may be inspired to do the same with their vision.

Kimberly Walton Jungkind

1

THE START OF A JOURNEY

Few are those who see with their own eyes
and feel with their own hearts.
—Albert Einstein

Surrounded by sun-drenched, rolling green fields with red country barns and a cluster of suburban homes, a young family struggled to care for a member who had chronic hepatitis B virus. Many families before them had faced this situation and searched for a solution. This family's search for a solution, however, would help them and many others around the world in the long run.

It all started in 1990, when a family's quest for information, through a number of connections, put them in contact with another Pennsylvania family. Dr. Timothy Block and his wife, Joan, deeply moved by the plight of this young family and their fight with chronic hepatitis B, wanted to help in some way. Dr. Block, an Assistant Professor and a research scientist at Jefferson Medical College in Philadelphia, knew that hepatitis B was a virus that causes a serious liver infection, which affects approximately one in three people worldwide.[2] Chronic hepatitis B infection can lead to cirrhosis or liver cancer.[3] With this information in mind, he understood the

---

2   Hepatits B Foundation website (December 12, 2011) http://www.hepb.org/hepb/
statistics.htm hepb.org
3   Timothy Block, personal interview, June 19, 2011.

dilemma for this young family and the impact of the hepatitis B virus.

While walking down the shady, tree-lined Delaware Canal tow-path where, long ago, mules were used to pull barges full of food and supplies through the canal, Dr. Block encountered his local New Hope friends and neighbors Jan and Paul Witte, who were also walking along the trail. Dr. Block spoke to the Wittes about this local family in need. "The Wittes are a very humani-tarian couple and were eager to lend a hand," recalls Dr. Block.[4] The Wittes, known for generously helping with community proj-ects and people, responded to this situation with compassion. The Blocks and Wittes agreed to help the Pennsylvania family in need, as a team effort.

The Hepatitis B Foundation founders: Dr. Block, Joan Block, Jan Witte and Paul Witte Reference: http://www.hepb.org/about/our_mission_and_story.htm

The team effort was truly a beginning. However, many questions required answers: How would they work together? What would be accomplished? What resources would be needed? What im-pact could they really have? Would their combined efforts make a difference? What would be necessary to make an impact?

They adopted as their own, the famous quote of Margaret Meade who said it best: *"Never doubt that a small group of thoughtful, committed citizens can change the world. Indeed, it is the only thing that ever has."*

The journey to help this family begins with the Blocks and the Wittes, and it was also spurred by the infusion of an "Albert Einstein–like" solution. Einstein and his wife traveled to America by steamship in October 1933 after he resigned from the Berlin Academy in Germany.[5] He and his wife arrived in New York and were driven to Princeton, New Jersey, and they quickly assimilated into a small home, just off the beautiful green lawns and stately buildings of Princeton University.

Einstein enjoyed meeting people and taking long walks in the park. On weekends, Einstein would travel a short distance away, with his newfound Princeton friends, to New Hope, Pennsylvania. He liked to walk along the banks of the Delaware River. Weekends away from Princeton were cherished; Einstein and friends would play cards and enjoy spending time in a guest cottage behind the larger home owned by a friend on the Delaware River.

More than seventy-five years later—in the same expanded guest cottage where Albert Einstein played cards on weekends—the Wittes and Blocks discussed what would be needed to help this young family. Ideas surfaced and meals were shared. More friends were contacted, and they joined the discussions. Providing this young family with helpful resources and understanding seemed to be the right thing to do.

Surrounded by creative artwork, the sounds of quiet music, and warm daylight hues of comfortable décor within the cottage, Einstein-like solutions began to emerge. Clinical expertise and awareness programs were needed, everyone agreed, along with specific education and probably long-term outreach for this long-term condition. Were there support groups or organizations available to help? Solutions, of course, required funding. How much funding would make a difference? What would be the focus? Would outreach and education be a part of the

---

5   Albert Einstein in Princeton. http://www.einstein-website.de/z_biography/prince-ton-e.html . (December 12, 2011)

plan? Researching to find a cure? This was a local tragedy as well as an international challenge. What were the next steps?

More dinners and lunches would be held to brainstorm ideas. An alliance comprised of the Blocks, the Wittes, and many others identified the need for resources, education, thoughtful energy, and much-needed research in an effort to contribute to curing chronic hepatitis B. From the start, the need for research was discussed as key and important to helping not only the young Pennsylvania family in question but other patients with chronic hepatitis B. Dr. Timothy (Tim) Block was eager to consider the laboratory-based research that was needed.

⌘　⌘　⌘

As in all efforts to change the world, leaders emerged. Timothy Block, PhD, raised in Buffalo, New York, was extraordinarily gifted. After reading a book called *Summer Hill* by A. S. Neill as a teen, Tim objected to the way current schools seemed to be "too regimented and conforming."[6] With a stirring sense of activism and in an effort to change this situation, Tim initiated, with friends, the creation of a charter school called the Ultimatheun "Free" High School, which employed volunteer faculty from the University of Buffalo as its instructors. The highly notable university professors were eager to teach and support these future leaders. After the charter school was off the ground with a small paid staff and faculty in place, Tim was accepted into the University of Buffalo at the age of fourteen after completing the New York Regents exam.

He studied biochemistry at State University of New York, in Buffalo. Dr. Block then completed his graduate and post-doctoral fellowship in biochemistry at Princeton University in New Jersey, and while there he also gained respect for a sport he had never played. "I actually played football as an extra-curricular sport when I was at Princeton,"[7] Dr. Block revealed. "I think it was a way to teach us teamwork as we were having fun and not trying

---

6   Timothy Block, personal interview, June 19, 2011.
7   Timothy Block, June 19, 2011.

to be competitive."[8] Dr. Block completed his graduate and post graduate training in Biochemistry at Princeton University in New Jersey. Shortly after graduating, his first professional position was in historic Philadelphia at Thomas Jefferson University. He joined the teaching and research staff as an assistant professor of microbiology and immunology, Jefferson Medical College. The same energy and creativity that had propelled Dr. Block forward in his teen years later enriched his scientific work life. With boundless enthusiasm and resourcefulness, he was eager to tackle this latest challenge of helping this family in need.

⌘　⌘　⌘

Building a successful structure and fund-raising were key priorities and the first steps to ensure the success of this endeavor. Later that year, in 1990, as the cold December snow blanketed the ground, a group of concerned people were invited to come together, brainstorm ideas, and potentially declare the start of a grass roots organization.

As warm pizza was served, the participants tried to settle on a possible name for the new organization. The New Hope Foundation seemed like a fitting and promising name. Then another name was thrown into the ring of choices that was bold and to the point: the Hepatitis B Foundation. "It was Paul's idea to start the foundation," Joan Block recalled. "Paul thought that it would be the best way to pull together the expertise and tackle the needs as a formal organization," said Joan.[9] They agreed on a foundation as the best structure for their efforts. Paul Witte, a creative and multi-talented professional, designed the logo, which was a bold HB with two semicircles buttressing the letters. This timeless logo would inspire many to help with the fledgling organization.

The pizza dinner was followed by other meetings and meals to draft the slogan for the foundation, "A cause for a cure." Of course, an attorney was contacted for legal clearance of the logo

---

8   Timothy Block, June 19, 2011.
9   Joan Block, personal interview, June 19, 2011.

and tagline, at which point the group celebrated! Joan was thrilled.

Joan Block, Tim's wife, was a co-founder of Hepatitis B Foundation. "Tim's biggest asset is Joan," notes Dr. Feldstein, interim president of Delaware Valley College, "and she is remarkable." Joan is a tremendous organizer, a caring advocate with a warm personality and a practical mindset. As a pediatric nurse, she understood the effects chronic disease had on a patient and the needs of the patient's family. Joan is a pediatric nurse, an active mom, and a kind-hearted person who springs into action to help anyone in need. She, along with her husband, Dr. Block and the Wittes, reached out to help the young family in need. There was some discussion about approaching other organizations for help.

A call was placed to the only non-profit organization that had *liver* in its name, the American Liver Foundation (ALF). That organization was not interested in a research arm that would focus on chronic hepatitis B.[10] The ALF was a growing organization with educational outreach as its focus. Accepting that the ALF was not interested in joining forces at that particular point in time, the fledgling grass roots group decided to continue to focus on their specific mission.

The enthusiasm grew in 1991 as Jan Witte confidently announced the securing of a post office box in New Hope for the exclusive use of the foundation. Tim Block and local entrepreneur Alison Kingsley prepared a comprehensive funding proposal of the foundation's objectives that would describe how a research facility and education would be the organization's hallmark. The need for seed money was apparent.

A committee was formed with Tim Block, Joan Block, Robin Larsen, Jamie Fox, and Alison Kingsley to plan a fund-raising event in the area. Jan and Paul Witte contacted Joey Luccaro, who offered his restaurant for a date in February. The plan was to raise money and find others who were interested in assisting

---

10   Timothy Block, June 19, 2011.

with the new organization. A voluntary board was needed and preparations to incorporate as a nonprofit were the next step.

As early 1991, as winter winds howled, the Wittes helped to organize a first fund-raising effort, which generated $17,000. The money was used for educational efforts. Jan described the event as "auspicious and mobilizing." "Our state senator (James Greenwood) and our U.S. Congressman (Peter Kostmayer)," Jan continued, "both delivered inspirational messages about the importance of bringing hepatitis B awareness to the public spotlight. The event drew a crowd of concerned and compassionate citizens who wanted to learn and wanted to know how they could help. We were overwhelmed by the response."[11]

Paul and Jan Witte, cofounders of the Hepatitis B Foundation, have always been leaders in their community. They worked tirelessly as a creative and caring team for the foundation. They were well-connected and understood the challenges and the tremendous effort that was needed to collaboratively work to find a cure for chronic hepatitis B. They warmly opened their house for fund-raisers for virtuous causes. Dr. Block describes the Wittes as having "selfless generosity."

Paul Witte, the president and founder of Originetics, Inc., is a product designer who specialized in the design of orthopedic products for twenty-three years until his retirement in 2000. Paul has received awards from *Industrial Design* magazine and the Museum of Modern Art. Jan Witte, a former elementary school teacher, worked tirelessly to establish the Hepatitis B Foundation and to keep the momentum moving forward. Jan is a full partner with Paul in their professional and philanthropic activities.

It was Jan who pointed out, in 1991, that New Hope, by its very name and nature, is the perfect place to start the nucleus of an organization. It is truly a network community that makes good use of the diverse professions and talents of the residents, especially when there is a worthy cause to support. Joan Block

---

11   Janine Witte, personal interview, October, 10, 2011.

describes Jan as "wonderfully compassionate and always responds directly and generously to an identified need."

Summer arrived, marked by warm breezes, rolling green hills, and fields of country flowers in Bucks County, Pennsylvania. A cold pesto pasta salad, grilled chicken and iced tea was served during a meeting at Einstein's kitchen table to decide the final list of Hepatitis B Foundation officers: Jan Witte, president; Alison Kingsley, vice president; Joan Block, secretary; and Jamie Fox, treasurer. The group agreed with Tim's suggestion to reach out to Baruch Blumberg in order to see if he would be willing to help the new organization.

Dr. Blumberg, who had received the Nobel Prize in Medicine in 1976 for his discovery of the hepatitis B virus, and later developed a vaccine against the disease, was eager to lend his expertise. An avid cyclist, former Master of Balliol College at the University of Oxford in the U.K., founding director of the NASA Astrobiology Institute and world traveler, Dr. Blumberg joined the Hepatitis B Foundation as its scientific advisor. He was pleased to become a part of the organization.[12]

The founders and supporters of the emerging organization were thrilled to have a Nobel laureate join their grass roots efforts. "The reason I got involved with the Hepatitis B Foundation," Dr. Blumberg recalls, "was because it had the promise of being an important center for hepatitis B research."[13] He also says, "It seemed like a formidable job at the time, starting from essentially scratch and [the Hepatitis B Foundation was] equipped with intelligence and energy and knowing how to get things done."[14]

Dr. Blumberg shared with the small group of new Hepatitis B Foundation officers more about the disease. Hepatitis B is defined as "an infectious illness caused by hepatitis B virus (HBV) which infects the liver and causes an inflammation called hepatitis."[15]

---

12   Baruch Blumberg, Personal interview. February 25, 2011.
13   Hepatitis B Foundation 20th Anniversary Video (December 11, 2011) http://www. youtube.com/watch?v=LUk2mPFs_cI
14   Hepatitis B Foundation 20th Anniversary Video (December 11, 2011)
15   The Free Dictionary, (November 21, 2011) http://medical-dictionary.thefreedic-

Originally known as "serum hepatitis," the disease has caused epidemics in parts of Asia and Africa, and it is endemic in China. About a quarter of the world's population, more than 2 billion people, have been infected with the hepatitis B virus.[16] This includes more than 350 million chronic carriers of the virus.[17] Chronic hepatitis B is the primary cause of liver cancer, which is the tenth leading cause of death in the world.[18]

As the founders learned more about the disease, their efforts prompted them to plan private educational receptions hosted by Gloria and Paul Blasenheim, Rita and Henry Lowe, and the Mattus family featuring Dr. Sanford Kuvin, an infectious disease specialist, who spoke about the hepatitis B. Another successful foundation community reception was held in Gladwynne with Dr. And Mrs. Forest Anthony, Mr. and Mrs. Ogelsby of Bryn Mawr, Nancy Evoy of Villanova, and Mr. and Mrs. Brengle of Haverford. Educational efforts continued with radio station interviews of Dr. Block by WDVR (89.7 FM) and a video narrated by Dr. Nancy Snyderman of ABC news.

Finally, brochures were created and printed to be distributed with the new quarterly newsletter: *Advice for HBV Carriers* and *Advice to the Parents of HBV Carrier Children*. These brochures were a great start to providing education on hepatitis B virus and were paid for by private donations from the public. Education was important, and the next step would be to appoint a board of directors for the foundation. Tim's brother, Tom Block, who is a car enthusiast and the co-founder and chief operating officer of Thomas Sweet Ice Cream and Chocolates, joined the board. John Keenan, MaryAnne McDonald, and Robert Miller rounded out the list of board members.

The newsletters were written by Jamie Fox and Jan Witte. The board would expand and become more specialized over the next few years, building on their foundation. Dr. Blumberg and

---

tionary.com/hepatitis+B

16   What is Hepatitis B (January 3, 2012) http://www.news-medical.net/health/What-is-Hepatitis-B.aspx

17   What is Hepatitis B (January 3, 2012)

18   About HBV Research Funding (December 23, 2011) http://www.hepb.org/about/hbv_research_funding.htm

Dr. Stanley Kuvin became members of the scientific advisory board and continued to provide their expert opinion to help the Hepatitis B Foundation

Research in Focus

With the board in place, it was time to direct laboratory resources and start the road to finding a cure for chronic hepatitis B. Research was paramount. The board hoped to bring hepatitis B to the forefront of research at Jefferson Medical College of Thomas Jefferson University (in Philadelphia) and to secure dedicated laboratory space for the foundation's use. Dr. Block, assistant professor and researcher at the time at Jefferson Medical College, working on herpes viral research. Tim reached out to Dr. Joseph Gonnella, dean of Jefferson Medical College, for advice on establishing a research center for hepatitis B there.

Dean Gonnella, an enthusiastic supporter, was receptive to the idea of providing research space. Numerous promising meetings and discussions about obtaining dedicated research space at Jefferson were held with Dr. Gonnella; John Monnier, senior associate dean; and other members of the leadership team at Jefferson Medical College. Conversations continued as research space needs were assessed by Dr. Gonnella to focus on hepatitis B research.

Within weeks, Dr. Block was granted permission to call his laboratory the Hepatitis B Foundation Laboratory. This was monumental and a great start for an emerging organization. The board of directors celebrated with joy and excitement! They now had dedicated laboratory space and equipment for researchers to make a difference.

It was also an acknowledgment of the value of finding out more about chronic hepatitis B. At that time, very few researchers in the world were dedicated to hepatitis B. On the third floor of a tree-lined historical Philadelphia center city neighborhood at a medical center, the research began.

Dr. Gonnella was supportive of the research and "was convinced that Tim would make a difference! He was bright, creative, enthusiastic, energetic and he believed in himself and the team which he had created. He was a dreamer, but because of his commitment I was convinced he would succeed." Truly, this was a significant step forward for the Hepatitis B Foundation, for hepatitis research, and for Jefferson Medical College.

The members of the Hepatitis B Foundation were thrilled that their laboratory would be located at Jefferson because of the medical college's nationally recognized reputation for scientific research. "This was an exciting beginning," said Dr. Block, "and I am committed to the cause for a cure." The new laboratory was approximately one thousand square feet, and the Hepatitis B Foundation was able to contribute $10,000 from various fundraisers to provide the means to start the research.

This early fund-raising contribution, along with the generous funding from Jefferson, allowed Dr. Block to hire Dr. Xuanyong Lu from Dr. Wolfram Gerlich's laboratory, and Mr. Anand Mehta to work as a research assistant on hopefully the next major research breakthrough!

At the same time, it became evident to the founders that a support group would be helpful for the initial family and for other families that were surfacing with questions about their own family members with chronic hepatitis B. In a corner office at the Fox Chase Cancer Institute, overlooking Burholme Park, W. Thomas London, M.D., director of the liver cancer prevention center, received a phone call from the Hepatitis B Foundation. Dr. London, one of the key researchers who had worked with Dr. Blumberg on the discovery of the hepatitis B virus, organized a support group with the help of Marianne Buzby, RN, a Nurse Practitioner from Children's Hospital of Philadelphia.

These support group sessions were very beneficial to patients and family members, who attended with numerous questions and desired information. Dr. London provided detailed answers and much-needed guidance, and he also agreed to become a member of the board. Dr. London had also been a member of the original team of laboratory scientists that assisted in the discovery of hepatitis B virus (HBV).

Scientific findings are sometimes serendipitous. Such was the case with the discovery of hepatitis B virus (HBV). Baruch Blumberg did not set out to discover the cause of "serum hepatitis," but an open mind and understanding of the scientific process led to a bull's-eye to which he initially had not been directed. As a medical anthropologist, Dr. Blumberg and his team were studying genetic differences in native populations by examining proteins in the serum of the patient's blood.

He realized that one of the serum proteins (which he named "Australian" antigen) turned out to be the hepatitis B virus when a patient they were observing tested positive for the Australian antigen and became sick with the clinical signs of hepatitis.[19] It was this process of discovery that eventually resulted in Dr. Blumberg receiving the 1976 Nobel Prize in Medicine.

A Collaborative Team

The long-term collaboration between Dr. Thomas London and Dr. Blumberg—epidemiological, clinical, and virological studies of hepatitis B and hepatocellular carcinoma (HCC)—has resulted in more than 265 publications written from the perspective of an epidemiologist, a clinician, and a virologist. Dr. London, who was recruited by Dr. Blumberg to join him at the Fox Chase Cancer Center (FCCC), initiated a liver cancer prevention program for Asian Americans living in the Philadelphia area in 1983 that continued until 2004.[20] The program combined screening for hepatitis B infection, immunization of susceptibles, management of chronically infected individuals, and early detection of HCC. For more than three decades, Dr. London has been a national leader in the effort to unveil the secrets of the hepatitis B virus and has specialized in researching how hepatitis B impacts the liver and how this viral infection actually leads to

---

19   Scientific process and the Hepatitis B virus. Creativity Research Journal, vol. 7, 1994.
20   Dr. Thomas London, Personal interview, February 14, 2011.

liver cancer. A thoughtful and caring physician, Dr. London was eager to lend his expertise to the Hepatitis B Foundation.

"The founders of the Hepatitis B Foundation were very kind people from the first time they contacted me," remembers Dr. London. He has continued to lend his expertise with numerous efforts on behalf of the Hepatitis B Foundation. Another important member of the collaborative team working with Dr. Blumberg at FCCC was Dr. Hie-Won L. Hann. Dr. Hann, formerly at FCCC, is a professor of medicine at Jefferson Medical College and the director of the Liver Disease Prevention Center at Thomas Jefferson University.

She has dedicated her life to the research, prevention, and treatment of chronic hepatitis B and has been caring for approximately three thousand patients on an ongoing basis.[21] A humble and distinguished expert trained by Dr. Blumberg, Dr. Hann has changed thousands of lives through her successful educational, screening, and treatment programs during her career. "Unfortunately there was limited therapies years ago that were available for patients with chronic hepatitis B," remembers Dr. Hann. "Now, there are seven approved medicines for patients with hepatitis."[22]

"The team of Drs. Blumberg, London, and Hann have been incredible advocates for Hepatitis B Foundation," reports Dr. Block. This was a strong and powerful team that would continue helping patients with for many years and leave a powerful legacy.

The powerful team of leaders in hepatitis from the Fox Chase Cancer Center became the scientific platform of consultation and thoughtful collaborative discussions that helped place the Hepatitis B Foundation on the launching pad as the next "rocket booster" of research. Just as NASA meticulously plans for its next launch with the right professionals, resources, and supplies, the Hepatitis B Foundation would need to successfully do the same.

---

21   Dr. Hie-Won L. Hann, Personal interview (August 25, 2011)
22   Hann, Personal interview (August 25, 2011)

The next stage was still to understand the key findings about HBV and somehow uncover answers that would hopefully lead to breakthrough discoveries of vaccines and medicines for people affected by chronic hepatitis B. The challenge was apparent, and it was a worthy endeavor.

2

RESEARCH AS A KEYSTONE

Anyone who has never made a mistake
has never tried anything new.
—Albert Einstein

As colorful autumn leaves covered the ground and pumpkins speckled the fields, more good news arrived. The Hepatitis B Foundation received its tax-exempt status, and the state had awarded the foundation a $10,000 grant under the PA Legislatives Initiative program. The award was contingent, however, upon matching funds of $5,000 in donations.

As the call went out for donations, the small foundation celebrated as it received funding from generous individuals and three small local companies. The funds would be used for two video projects, which featured experts on hepatitis B as well as people affected by the virus. Hepatitis B Foundation was also able to produce, thanks to a state grant two additional public service announcements for television.

Soon, New Hope mayor James McGill was assisting with promoting fund-raisers for the Hepatitis B Foundation. One fund-raiser was the New Hope Performing Arts Festival's "Men 'n' Dogs," a production of a one-act comedy by Peter Tolan. Congressman Peter Kostmayer hosted eighty people at a fundraising event that included a guest speaker, hepatologist Dr. Friedman from Thomas

Jefferson University. Cofounder Paul Witte stated, "The [event] provided the foundation with funds to pursue other avenues of funding that we are confident will generate the initial capital to finance specific research projects in finding a cure for hepatitis B."[23]

The Oxford Sabbatical

Of course, the foundation knew that the real win would be to find a cure for chronic hepatitis B. A carefully planned, concerted research effort would need to be the next and most important focus. With his professional awards from the American Cancer Society and the International Union Against Cancer, Dr. Block knew that he needed to work and collaborate with the best researchers devoted to hepatitis B.

In September 1992, he decided to take a sabbatical leave from Jefferson Medical College to further his current research. He believed that HBV proteins and the development of an anti-hepatitis assay might help determine the best drugs for treatment of hepatitis B. Dr. Block would spend the next year working under the guidance of Dr. Blumberg in Oxford, England, to test the possibility that certain compounds could become incorporated into the HBV structure and disable the virus.

Dr. Block was thrilled to have the opportunity to work with Dr. Blumberg, who was an inspiring mentor. Dr. Block's research would be conducted in the laboratory of Raymond Dwek, a foremost authority on glycoprotein structure and the director of the Glycobiology Institute at the University of Oxford.

Before Dr. Block left for England, however, he attended the International HBV conference held in San Diego, California, at the end of the long, hot summer of 1992. At the conference, Dr. Block had the opportunity to report on the use of his "designer bacteria" to search for and generate novel anti-viral compounds. The presentation and discussion at this meeting marked a promising prototype. Unfortunately, only a few of the papers presented at the meeting dealt with research related to the HBV cure,

---

23   New Hope Gazette, February, 28,1991.

which was only too similar to the previous year's conference. It was clear that the quest for a cure was yet to be completed.

Drs. Timothy Block and Barry Blumberg reviewing data.

Off to the UK! Arriving on a damp, cold, and rainy day in Oxford, England, Dr. Block, Joan, and their two children organized their new temporary home. It was a small, cozy duplex that was located in a Cotswold-type village about five miles from Oxford. Life in England was similar to home in many ways, but also different. Women in this town would shop daily at the butcher, baker and green grocer. They would also visit each other and share tea while their small children played. They grew accustomed to "popping by" the local pubs in the evenings, when families gathered to socialize. They also experienced formal gatherings at which men wore morning glory (long) suit coats as a part of the British tradition. The Blocks were honored to be invited to sit at the "high table" and enjoy delicious dinner fare at the University of Oxford events.

On one formal, shady, outdoor occasion at Oxford, Dr. Block was dressed in formal attire and was intermittently pelted with wine bottle corks thrown by other attendees when he took off the long coat because it was so dreadfully hot. The Oxford custom, he learned, was to leave the coat on even if it was uncomfortably hot. Endeavoring to follow tradition, Dr. Block donned his coat with tails and enjoyed the event.

The University of Oxford is the second-oldest surviving university in the world. It dates back to the eleventh century. The university is a federation of more than forty self-governing colleges and halls, along with a central administration headed by the vice-chancellor. More than forty Nobel laureates have been associated with the university, and its list of distinguished scholars is long and includes many who have made major contributions to British politics, the sciences, medicine, and literature.[24] The stately medieval buildings connected with arched walkways and manicured green lawns underscore its reputation as the top university in the United Kingdom.

As a sabbatical fellow in Balliol College at Oxford University, Dr. Block worked closely with other scientists, such as Royal Society member Professor Raymond Dwek and Nobel laureate Dr. Baruch Blumberg. Together they discovered a new mechanism in the replication cycle of hepatitis B virus that could be a clue to the treatment of the disease. Dr. Dwek, often called the father of modern glycobiology, remembered Tim as a "star" and "full of energy." He added, "Tim was a popular professor with everyone in the lab and had a wicked [great] sense of humor."[25] As the months went by, the HBV research team in Oxford blossomed into a strong collaboration.

Breakthrough Research

It was in this laboratory that NBDNJ, a compound that showed promising activity against HBV, was identified by Dr. Block. Dr. Block discovered this plant sugar which has now received the trade name *glucovir*™. To summarize the research finding, HBV grows inside liver cells, and after infecting these cells the virus replicates and "buds" to the outside, where it enters the bloodstream and attacks other cells.[26] The virus can be "locked" inside the infected cell by specific sugars. The sugars surround the cell in this case, preventing it from replicating. Dr. Dwek described the HBV research as "breakthrough" and used the

---

24   University of Oxford.(January 10, 2012) http://www.ox.ac.uk/about_the_university/oxford_people/oxonian_award_winners/index.html

25   Raymond Dwek, Interview via phone, (April 24, 2011)

26   Dwek, Interview via phone, April 24, 2011.

analogy of the hepatitis B virus as a "football that is punctured" and cannot replicate.

Glucovirs™ inhibit glycan processing, and its mechanism of action provided some of the first insights into protein folding (Block, Dwek, Blumberg), some of which are often cited as the definitive work in the field. However, perhaps of greatest significance is one of the drugs that Dr. Block co-discovered: Alkovir™ 231, developed by United Therapeutics, Inc., is in phase three clinical trials in the United States for treatment of hepatitis. Another, developed independently by Enzo Pharmaceuticals and Oragen, Inc., called EHT 999 is entering phase three human clinical trials.[27] It is hoped that these discoveries will lead to medicines that can help people with hepatitis B in the future. This was exciting news for the world of hepatitis B.

The leadership of the Hepatitis B Foundation in Doylestown and in Oxford continued to track newsworthy research around the world on hepatitis. Texas researchers, for example, believed that a hepatitis vaccine could possibly be manufactured from garden vegetables. Scientists found that after putting hepatitis antigen into a solid virus, vegetable plants infected with the virus began producing small amounts of antibodies.

FAST FACT:
This year alone, 100,000 new people will become infected with HBV.

Other research at the Fox Chase Cancer Center in Philadelphia reported that a drug called carobodeoxyguanisine (CDG) was effective in cleaning 99 percent of the duck HBV from chronically infected ducks following a single treatment period. Unfortunately, the virus eventually bounced back when the drug treatment was stopped. Another scientific research team from Innovir Labs, Inc., demonstrated that ribosomes can be designed to specifically cut and destroy HBV RNA. Ribosomes are sometimes referred to as molecular scissors, as they can be designed to cleave at a particular point, allowing a drug to attack a virus. Every one of

---

27   Timothy Block, personal correspondence: Curriculum Vitae June 6, 2011.

these small findings are important parts of the puzzle and may hold the key to unlocking the cure for chronic hepatitis B.

The big therapeutic news, at that point in time, was that interferons were effective for some individuals based on the work done by Jay Hoofnagel and other researchers.[28] Other news included the understanding of how HBV replicates and that this may lead to the discovery of drugs that suppress the enzyme that HBV uses to replicate.[29] Treatment breakthoughts came in the late 1990's by scientists around the country!

Researchers in any scientific subject make a habit of monitoring the findings of others who are concentrating on the same general topic. They share their efforts with other researchers in an effort to further the science and build on each other's work to get them closer to their own research goals. Open collaboration and published work of researchers around the globe in different journals was and is vital to gaining nuances as well as a broader perspective of what is possible and what has already been noted as an accomplishment.

By the spring of 1993, researchers noted that preliminary tissue culture studies of a particular compound appeared to profoundly influence the assembly of HBV. The compound apparently causes the virus to develop aberrant sugar residues on its surface. Although HBV normally has sugars on its surface, the presence of these aberrant sugars causes it to get stuck and stay in the infected cell or fall apart once it leaves the cell.

In an effort to reproduce these findings in another lab to make sure he had the same outcome, Dr. Block traveled to Germany for two weeks to test the experiments in one of the world's leading HBV molecular biology labs. The data was reproduced in Germany. The results were encouraging, and the finding represented a piece to the larger puzzle of finding a cure.

---

28   "Dr. Jay Hoofnagle Honored by HBF" Hepatitis B Foundation http://www.hepb.org/about/hufnagle.htm (April, 2003)

29   HBF Presents Distinguished Scientist Award, http://www.hepb.org/pdf/Gala_2006_press_release.pdf (March, 2006)

Growing Momentum in Pennsylvania

Meanwhile, across the pond in Doylestown, Pennsylvania, a small office was rented for the Hepatitis B Foundation. The organization continued to strengthen regionally as contacts were made with hepatologists, gastroenterologists, public health professionals and community leaders. In order to help the number of patients contacting the foundation for help, a Hot Sheet was created of useful phone numbers of doctors, clinics and other healthcare professionals. Increasingly, the foundation was serving as a gateway for those interested in information and support related to hepatitis B. To that end, an outreach and education plan was drafted and implemented. quarterly newsletter.

A quarterly newsletter, called *B Informed,* was a way to disseminate accurate information about the Hepatitis B Foundation to patients, families, friends, and healthcare professionals. The first issue informed readers about a new fund-raising idea for the Hepatitis B Foundation. A series of "cabaret nights," with cocktails, gourmet dinners, and songs and theater, became the talk of the town.

Swaying to Gershwin music and clinking glasses were the sounds of these joyous events. These cabaret nights were followed up by "Sundaes in Solebury." Friends enjoyed large ice cream sundaes piled high with Thomas Sweet Ice Cream from a delicious local creamery. The old fashioned party atmospheres of these gatherings were fun and well-attended. Everyone had a wonderful time supporting the Hepatitis B Foundation and enjoying these fund-raising events.

The Hunt for a New Research Home

These events were quite important as enthusiasm grew for the next level of commitment: building a new research center where the focus of research would be to find a cure for hepatitis B and the search to find a cure for this devastating disease. As Dr. Block, Assistant Professor of Microbiology and Immunology at Jefferson Medical School in Philadelphia, PA returned to the United States, in the Fall of 1993, with research enthusiasm and the growing support of local friends and advocates, the idea took shape through conversation about the growing need to establish a larger Jefferson laboratory space with more financial resources dedicated to hepatitis B.

Dr. Block spoke again to Dr. Gonnella, dean of the medical school. Under the dean's leadership, Jefferson Medical College generously offered a $1.5 million match of in-kind support to be used over a five-year period for HBV research at Jefferson. "I saw Tim's enthusiasm," recalled Dr. Gonnella, and "wanted to help his with his goals." The enhanced research laboratory was officially opened in 1995 and titled The Foundation Laboratory at Jefferson. Celebrations were underway as meetings took place at the new Hepatitis B Foundation office in Jenkintown that opened in 1996 near the train station.

Research was underway with a small group of energized laboratorians on the third floor in the Alumni Building of Jefferson Medical School. Unfortunately, enthusiasm was somewhat dampened within a year or so when it became apparent that more equipment was needed and the current space was actually too small. Dr. Block described the idea of an enlarged research center for hepatitis B. This center would, in Dr. Blocks words, "concentrate researchers as an incubator facility, and it would be a family of scientists that would increase the intellectual universe" of the research. Dean Gonnella was briefed on the idea and was very supportive. Dr. Block set out to find available larger research space in the surrounding region.

A local Naval Air Warfare Center in Warminster, Pennsylvania, was in the process of realignment, and there was available space for a lab. The site was acquired by the U.S. Navy during World War II from the Brewster Aircraft Corporation; it served as a strategic locale for weapons development and testing of modern aircraft. Later, it was a training center for America's Mercury, Gemini, and Apollo space programs. Toward the end of World War II, it grew to become one of the leading naval air research laboratories in the nation. It was the home to one of the first and largest human centrifuges, which was used extensively for astronaut training during Mercury missions.[30]

---

30   NAWC/NADC Warminster Historical Information. http://www.navairdevcen.org

Dr. Block toured the facility with Donald Jungkind, PhD, professor in the microbiology and pathology department at Thomas Jefferson University Hospital, and Jim Greenwood, United States Congressman. Dr. Block recalls, "We were taken very seriously at the navy base and discussed transferring the lab to this location as a multi-use facility with other interested companies." The base was impressive, with plenty of space to convert large storage facilities and buildings into a grand research center. This potential site had the advantage of federal tax breaks for interested parties requiring expansive space for a growing business.

Another location under consideration offered tax breaks as well. The University of Delaware was also interested in having a state of the art research and classroom facility for its students. Dr. Block noted that the "college leadership met with me to discuss the research lab idea," and that university leadership "were all intrigued, especially the University of Delaware's college-dominated town."

Both the Naval Air Warfare Center in Warminster and the University of Delaware seemed like viable options for development of a research facility for hepatitis. A third option surfaced when Dr. Block's mother called and said, "Call Delaware Valley College and speak to the leadership—it's a nice college and it's right here in Doylestown."

Delaware Valley College (DVC) was established in 1896 by Rabbi Joseph Krauskopf, DD, who purchased a hundred-acre farm, arranged for the construction of a small classroom building, employed a faculty of two, and enrolled six students. DVC, then called the National Farm School, provided a three-year program combining academics and agricultural experience that continued through World War II.[31] The college had expanded since then, and in 1969 it was officially renamed Delaware Valley College.

Dr. Joshua Feldstein, who was currently the interim president of Delaware Valley College, was contacted by Dr. Block and was

(January 21, 2012)

31   Delaware Valley College web site: http://www.delval.edu/pages/admissions/C96/
Accessed: November 10, 2011.

thrilled to provide a personal tour of the academic/agricultural campus with planted fields and stone buildings. Dr. Feldstein was interested in discussing the creation of a research center with laboratories, offices, and classrooms connected to the college.

He was also intrigued to hear about one of the researchers that Dr. Block wanted to bring, a professor named Dr. Hillary Koprowski, who was experimenting with plant-derived vaccines. He theorized that genetically altered, greenhouse-grown tomatoes were the optimal plant-derived way to deliver an oral vaccine. Because of this agricultural aspect to their research, building a highly sophisticated greenhouse at the new research center was a necessary part of the plan.[32]

Dr. Feldstein persuaded to the Delaware Valley College board to support the idea of a research center on campus. Dr. Block was eager to meet with Dean Gonnella and discuss the location as an option. "Dr. Gonnella was a wonderful man," Dr. Block recalled. "Dr. Gonnella went out on a limb for me [in support of this idea]." He understood Dr. Block's enthusiasm for expanding the current space and opening a separate research center in Doylestown.

Dr. Gonnella continued by saying that "Tim was so enthusiastic, full of energy, and on the go—he was such a good teacher." It was clear that Dr. Gonnella and other leaders at Jefferson took a risk in building a research laboratory in Doylestown as a part of Delaware Valley College. Many factors were considered and many meetings took place with experts at both institutions to iron out the details.

A New Research Home

"Write up a plan that would be good for Jefferson and Delaware Valley College, and I will get behind you," Dr. Gonnella advised Dr. Block. Dr. Block drafted a proposal that included laboratories, classrooms, office space, conference room space, etc. In the end, Block added, "This will be an exponential lift for science and research for Delaware Valley College and Jefferson." The proposal, however, needed to be reviewed by John Monnier, dean of finance

---

32   "The Tomoto Vaccine" Time magazine (November 18, 2002) http://www.time.com/time/magazine/article/0,9171,1003703,00.html#ixzz1cDiDhPUR

for Jefferson Medical School. Dean Gonnella, Dean John Monnier, and Dr. Block all worked together as a close-knit team to find the right balance of resources and structure a final proposal that benefited all parties. John Monnier agreed to the initial plans, but he worried that "there was a financial risk involved which needed to be considered ... especially if Jefferson guarantees the lease."

The final proposal, entitled "The Science Research Center in Bucks County," was described by Dr. Block as an "unprecedented philanthropic opportunity to create an original, world-class science research center for the Hepatitis B Foundation."

The "innovative, multi-user, nonprofit, public and private complex," as Dr. Block described it, "was a leafy liberal Eden research facility designed to be built on the Delaware Valley College campus." The unifying theme of the center was to be "translational research," and the unique feature was the combination of academic/basic science labs and a greenhouse next to the facility. The labs would include those of the Hepatitis B Foundation, Thomas Jefferson University, and other similar organizations with emerging technology, including biotechnology companies, some of which were working in conjunction with the Ben Franklin Centers of Pennsylvania.

Dr. Block commented that Dr. Feldstein "was very excited, committed, and confident" about the building and the opportunity for Delaware Valley College to have access to cutting-edge biotech researchers on the campus. The expectation was that the "Hepatitis B Foundation would locate enormous talent that was a mission-oriented research program," said Dr. Block, and "we needed Delaware Valley College to build us a building."

Dr. Feldstein said, "What a wonderful project," and "Can you be back up here tomorrow?" Dr. Block was enthusiastic, and in 1996, the building was built and paid for by Delaware Valley College, with a ten-year lease (1996–2006) to Jefferson, with the Hepatitis B Foundation as a subtenant. The lease was guaranteed by Jefferson Medical College.

"I did not want to not want to lose Tim, because he had a vision so I eventually gave him my blessing [to move forward]," recalls

Dr. Gonnella. The new three-story center was named the Jefferson Center for Biomedical Research and would include more than fourteen thousand square feet of space for the Hepatitis B Foundation, an office suite with a conference room, student classrooms, and two floors of laboratories, on a beautiful green grass and flower-adorned campus. It even included the first elevator on campus! Behind the building were acres of fields where you can always see farm equipment or horses cared for by the students.

The cost was slightly higher than anticipated. "Even though the rents from TJU to Delaware Valley College were intended to be sufficient to cover the full expense of the new building construction, the payments did not cover all of the costs," remarked Dr. Block. Parties connected with the new construction thought that it would be cost-neutral, and that was the intention.

"I believe Jefferson was motivated to help Hepatitis B Foundation and to retain me and because of the leadership of Joseph Gonnella, TJU's dean," said Dr. Block. He continued, "For Delaware Valley College, it was the vision of Josh Feldstein, Delaware Valley College's president, who recognized that the arrangement brought professional scholarly research science and outreach to the Delaware Valley College campus."

Dr. Feldstein noted that "accomplishments take vision, visionary people, and people dedicated to that vision." He also remarked that Joan Block "was an excellent executive director, and her commitment and contribution as the executive director for this new center was very remarkable."

From its humble beginnings at Einstein's guest house, the Hepatitis B Foundation could now call this state of the art facility its own. The members looked forward to taking advantage of the environment and adequate space to conduct world class serious collaborative research and an office to coordinate outreach materials and become the "go-to" place for hepatitis B research and education. Arrangements were made to move from the current small and cramped Doylestown office to the Delaware Valley College campus offices.

Joan Block, the first executive director of the foundation, was active in continuing to create the *B Informed* newsletters, answering the phone and managing the busy office almost single-handedly. In 1998, the Hepatitis B Foundation offices and labs were finally completed at the new Jefferson Center for Biomedical Research in Doylestown, Pennsylvania. The structure was welcoming with a large exposed glass front atrium to greet visitors with a stone exterior.

The Hepatitis B Foundation's Jenkintown office was moved to the new center at the Delaware Valley College without missing an incoming call from a patient or family member. The number of board members expanded, even though the Jenkintown office could barely hold Board meetings as it was sharing space with another non-profit organization. The foundation staff would not miss the small office quarters in Jenkintown. With research continuing, educational outreach was the next area of focus.

The Hepatitis B Foundation welcomed Lisa Honig, the new outreach coordinator, to lead the growing need for activities. Several outreach efforts were initiated.

- A support group of adult carriers was started for the Philadelphia area.
- A thirty-minute video was created, with an introduction by ABC TV's Dr. Nancy Snyderman.
- The foundation began to participate in local health fairs.
- A hepatitis B educational program was adapted for schools.
- The popular Sundaes in Solebury continued.

Dr. Donald Jungkind, who led the Laboratory Biosafety Committee at Jefferson, was asked to inspect the new Jefferson Center at Delaware Valley College as it was coming to completion to make sure that it met all of the biosafety level 2–3 standards for safe working practices. During that inspection process, Dr. Jungkind and his son Daniel invited Dr. Block and his son Peter to a weekend outdoor adventure at a local state park.

Woodland Adventure

Dr. Block was always open to an interesting expedition and agreed to join in on the father and son outdoor adventure weekend. They packed up a VW camper van and drove to Jim Thorpe State Park. As they jumped on their bikes and began pedaling alongside the Lehigh River, they steered around what they initially thought was a stick. They later realized it was actually an eastern Pennsylvanian black poisonous rattlesnake! Dr. Jungkind's son Dan took out his slingshot and hurled a few stones at the snake, which curled up, rattled menacingly, and eventually slithered off up the mountain.

That night, at a dinner campfire cookout, another uninvited visitor arrived. The adventurers had left a bag containing some food on the picnic table after dinner. Attracted by the scent, a forest critter crawled into the bag and began eating its contents. The fathers and sons couldn't see the opening of the bag from inside the VW camper van. As they watched the moving bag, they wondered what kind of critter was inside.

After a while, the critter finally came out and retreated from the campsite, at which point they were able to see that it was actually a skunk. For the sake of safety, the fathers and sons gathered their sleeping bags and pillows and jumped into the four-person sleeping van for a night's rest. The next morning, they scraped enough food together for a morning breakfast and drove home. Dr. Jungkind reported, "Everyone had had enough adventure for one weekend!"

Meanwhile, there was much excitement at the new Jefferson Center offices too: in an effort to meet the demanding needs of the foundation's outreach and educational efforts, the staff expanded to include Molli Conti and Peggy Farley. Molli recalls, "Dr. Block never stops thinking about the next step for both his research team and Hepatitis B Foundation." The vision for the new Jefferson Center building was to create a place where patients, families, physicians, and researchers would have access to information and excellent science rooted in human compassion.

3

PATIENTS AND
FAMILIES AT THE CENTER

Truth is what stands the test of experience.
—Albert Einstein

"Some patient calls were heartbreaking," recalls Peggy Farley, Director of Outreach Programs. Educational outreach and providing support to callers continued to be the focus. "I was completing my marketing degree at DVC and Mollie Conti, executive director of Hepatitis B Foundation, asked me to join the Hepatitis B Foundation staff." The "consult line" was very active at the foundation.

"We get all kinds of calls to the center and then send out brochures as a follow up to the phone conversation," says Peggy. There were also a lot of visitors to the Hepatitis B Foundation office, including local politicians, researchers, community leaders, students, and so many others interested in the mission of curing hepatitis B. To meet the demand, the office staff expanded and quickly settled into the Hepatitis B Foundation offices, conference room, and the reception area.

To reach a wider public and professional international audience, Mollie Conti, director of programs for the Hepatitis B Foundation, and Chari Cohen wrote a grant, and the foundation was awarded a three-year NIH grant of $450,000 to completely update the website. The goal was for the Hepatitis B Foundation to be the main on-line portal for hepatitis B information worldwide.

Chari reports, "We reached our goal, going from about two thousand on-line visits a year with the old BF website to more than 1 million visits from eighty countries a year with the revised website.

The website would be available in different languages, including Chinese, Korean, Spanish, Vietnamese, and Turkish, and it would contain information about drugs, education, liver specialists, clinical trials, and more. Once the site was completed, the foundation was flooded with e-mails from around the world, and calls to the office increased.

The staff kept in mind that patients counted on the Hepatitis B Foundation as a credible resource of accurate information. Patients and families learn that the number of persons living with chronic hepatitis B in the United States, according to the CDC, is about 800,000 to 1.4 million people.[33] The number of chronic liver disease deaths associated with viral hepatitis B each year is about 3,000.[34]

The challenge taken on by the Hepatitis B Foundation is to find a cure and to help people who live with an increased risk of dying prematurely from cirrhosis, failure and liver cancer. Patients and their families contact the foundation for information and emotional support. The following true story was originally published in the spring 1999 issue of the *B Informed* newsletter and represents many of the stories that the staff respond to on a daily basis.

Living with a Six-Year-Old—A Mother's Story

About two years ago my husband and I decided to adopt two children from Russia. We had hoped for a brother and sister under the age of five years. We knew that this

---

33   CDC.gov. September 4, 2011 http://www.cdc.gov/hepatitis/Statistics/index.htm
34   http://www.cdc.gov/hepatitis/Statistics/index.htm

would be a fairly easy match, but we were still surprised when our wish came true just seven weeks after signing on with our agency. The medical information we received was very scarce, but it did include the fact that our son had hepatitis B.

I knew very little about hepatitis, just that school had required our two older children to have the hepatitis B vaccine and that the disease could be fatal. We began researching the illness, contacting local doctors and also the Hepatitis B Foundation. Adoption is often described as a leap of faith; you have little or no knowledge of the children that you are about to bring into your family.

In the last several months, I have joined a group of other moms of children with hepatitis B. Our small but growing group is now eight and, as it turns out, all of our children have been adopted from various countries. In most cases, the discovery of hepatitis is made after returning home. Parenting a child with hepatitis B, I have found parenting a child with hepatitis B to be emotional at times, yet just like our three other children, our son is a blessing from God.

I have been fortunate to have been able to find help, assistance, and friendship from many sources. Hepatitis can be very frightening, but educating oneself, family, and close friends helps to separate fact from fiction. Hopefully, through groups like the Hepatitis B Information and Support List, people affected by hepatitis B can find strength from one another.

My son is a charming little boy with medium brown hair and big brown eyes. Even though he has hepatitis B, it is important to remember that he is a boy. He goes to school, rides his bike, takes swimming lessons in the summer. When he grows up, he wants to play baseball. My first responsibility to him is to be his mom and to love him. Although I continue to do research, looking at new drug trials and hoping for a cure, I try to focus on the little boy, not the disease.[35]

This mother reached out to the Hepatitis B Foundation for information and tried to understand how the disease was transmitted to her child. Hepatitis B can be transmitted through exposure to infected blood, unprotected sex, unsterilized needles, (ie. illegal drugs and medical or dental procedures), and from an infected pregnant woman to her newborn during the delivery process.

About Hepatitis B Virus Infection

Hepatitis B, however, is not transmitted through casual contact such as sharing food drinks, holding hands, or hugs and social kisses. In many cases, a person with HBV may not know he or she has the disease unless clinical symptoms are present or the patient knows of other family members with the illness and takes the opportunity to be evaluated by a physician and tested for the disease. Persons with chronic HBV infection may have no evidence of liver disease or may have a spectrum of disease ranging from chronic hepatitis to cirrhosis or liver cancer.

The HBV chronic infection may also be asymptomatic.[36] Approximately 0.3 percent to 0.5 percent of U.S. residents are chronically infected with HBV; 47 percent to 70 percent of these persons were born in other countries.[37] Regions of the world with high or intermediate prevalence of the hepatitis B virus by the presence of HBV in the blood include much of Eastern Europe, Asia, Africa, the Middle East, and the Pacific Islands.[38]

"Over the years," says Joan Block, a professional nurse, "we have heard from thousands of people who have been infected with hepatitis B and from parents who discover their child has hepatitis B." She continues by saying "The stories usually begin with feelings of confusion and anxiety in their quest for information, or in the case of their parents, too often the stories are about

---

35 Excerpt from B Informed, pg. 4. (Spring, 1999)

36 http://www.cdc.gov/osels/ph_surveillance/nndss/casedef/hepatitisb2007.htm accessed Sept. 4, 2011

37 http://www.cdc.gov/mmwr/preview/mmwrhtml/rr5708a1.htm

38 http://www.cdc.gov/mmwr/preview/mmwrhtml/rr5708a1.htm

their child's rejection from schools or other family members and friends. All of their experiences are very personal, sad, poignant, or sometimes even victorious, but each and every story is a daily reminder that what we are doing is important and that Hepatitis B Foundation meets a very human need."[39]

Recent stories have surfaced about medical and dental students who have not been accepted to U.S.–based medical and dental schools after being identified as hepatitis B positive. Members of the Hepatitis B Foundation board and the scientific advisory board have contacted the leadership of these universities to advocate for affected students. In some cases, sharing information about hepatitis B has been helpful.

In other cases, a medical or dental school may take a restrictive stance and not permit to attend or may be offered alternative graduate studies. Physicians and nurses, who are or become HBV positive and turn into chronic carriers, may be asked to pursue alternative clinical work. For example, a scrub nurse working in an operating room may be asked to become a circulating nurse. The foundation has been counseling students and clinicians in similar situations for many years.

The following is an excerpt from a story that was published in *B Informed* that highlights the challenges that patient's face who have chronic hepatitis B related to employment:

Carl's Job

> Carl is a salesman that worked in Ohio for seven years. He had kept his chronic hepatitis B a secret at work, but the time came when he felt he had to tell his boss about his situation. Increasing fatigue caused him to miss work occasionally and his blood test reports indicated that his body had reached an ideal condition to start rigorous treatment. His hepatologist recommended that Carl begin six months of treatment.

---

39   Standing on the Edge of Tomorrow, B Informed, Autumn, 1997, p.2

Carl went to his supervisor well-prepared to explain how the disease and the treatment might temporarily impact his work. After the discussion with his supervisor, two weeks later Carl received a letter of termination from his company.

Carl's story illustrates the dilemma that "hepBers" face at work. On one hand Carl decided to disclose his diagnosis of Hepatitis B. On the other hand, it may open up a legal situation and a review of current laws with the accompanying stress.

In this situation, Carl decided to settle with his ex-employer out of court. He had quickly found a new job, so money wasn't as much a motivating factor as the emotional blow of suddenly going from "valued employee" to "pariah." But he insisted on one stipulation: that the money from his settlement would be donated directly by his employer to the Hepatitis B Foundation.[40]

Joan Block understands the challenges that students and employees face and states, "It is very gratifying to all of us at Hepatitis B Foundation to feel that we can make a difference—by lending a knowledgeable yet sympathetic ear, providing accurate and timely information, and reassuring both adults and parents of children that the future for chronic hepatitis B carriers is very hopeful since several new treatments have appeared on the horizon."

"Even as more therapies are available, the issue of disclosure is most troubling to patients, because the stigma of having a diagnosis of hepatitis B," underscores Peggy Farley, community relations manager of the Hepatitis B Foundation. Patients' fear of revealing their own history, their family history, or current infection can become a barrier to treatment. She continues, "It is important to find out the status of your hepatitis B infection even if you are unable to get medical care right away or do not require immediate medical attention."

---

40   Carl's Job. (Excerpt) B Informed, Winter 2005, p.13.

Over time, this would ideally include annual monitoring of your HBV status. It is also important to protect your loved ones from acquiring HBV by ensuring they are screened and then vaccinated against HBV using the safe and effective three-vaccination series. Unfortunately, with the rising costs of health care and many without jobs or health insurance, getting care can be challenging, but there are options..." Peggy adds. She also suggests calling your local or state department of health.

Ask if there are free clinics or health-care centers to receive testing and care. Sexual partners and close household contacts of infected individuals can inquire about free screenings and/or vaccination clinics. The HBV vaccine is only effective if you are *not* already infected, so screening is important if close contacts have been exposed. There may be a Federally Qualified Health Center (FQHC), local university medical center, or state adult hepatitis B coordinator who may know where you can get tested.

Once you know the status of your hepatitis B infection, you can take some basic steps to improve your general health and, equally important, your liver health. If you smoke, quitting is recommended. Drinking is particularly damaging to the liver and should not continue. Maintain a healthy diet and get plenty of rest and moderate exercise. This may seem simple, but it is important. Of course, the Hepatitis B Foundation staff have been available to help provide more information and referrals to liver specialists (Hepatologists).

Children may be able to receive vaccinations free of charge via the Vaccines for Children Fund. Refer to Patient Assistant programs in Chapter 8. Qualified adults can be vaccinated using the Glaxo Smith Kline (GSK) Vaccines Access Program. (http://www.gsk-vap.com/) Pregnant women should be screened during pregnancy or prior to delivery for HBV infection. Hospital protocol should include the vaccination of all newborns within twelve hours of birth to ensure that they are protected from an HBV infection.

Stories of children living with hepatitis B are heard regularly by the foundation staff. The following story has been published in the *B Informed* newsletter and underscores the challenges for parents of affected children.

XiXi Comes Home

Club feet or cleft palate? Cerebral palsy or chronic heart disease? HIV or HBV? Or any combination of the above? For a prospective adoptive parent, a special needs checklist includes conditions no parent ever wants their child to have.

With our first adoption, we checked the box for hepatitis B, confident that it was a need we could handle. Our wonderful daughter Maya came home from China at eleven months. At age three, she began a yearlong treatment with pegylated interferon, resulting in a loss of hepatitis B surface antigen and the blessed addition of surface antibodies.

Not long after Maya finished treatment, a little boy's picture captured our hearts. He was three years old, sitting on a pink bike, flashing a smile as large as the Yunnan Province, his fingers raised in the "peace" sign. I read his file and almost cried: his special need was hepatitis B.

The smell of alcohol wipes was fresh in my memory. The syringes, the sharps container—I was done with that. I did not want to see another child go through treatment, knowing there were no guarantees. And yet, we couldn't forget this boy's picture.

I contacted our adoption agency and asked if other families were considering this child. No, they said, there were no other families. Gulp. "Well, we haven't done any paperwork," I said, sure that that would disqualify us, as China requires a family to have a completed file before selecting a specific child. Our agency said they'd call their contact in China. I left for errands, positive that our request would be rejected. I came home to the following message: "The boy's file has been available for over five months with very little interest ... non-infant boys with hepatitis B are not typically adopted quickly.... China has approved your request and we've locked his file." I almost fainted. We were heading back to China.

We met our son, XiXi, eight months later. He was huddled in a corner, afraid and near tears, clinging to the photo book we'd sent to the orphanage. A woman gently pushed him toward us, and he burst out into loud sobs. His sister, Maya, who couldn't remember her own similar experience, rubbed his back and repeated the Mandarin words of comfort that we taught her.

When he calmed down, we sat on the couch as he looked through his photo book. He pointed to a picture of himself as an infant and asked, "Shurbooshur wo?" ("Is that me?")

"Shur," I said. "Yes, that's you." It was him from the very beginning. It was him in China. It would be him in the United States. It was him with or without a special need.

And the hepatitis? His doctor personally called us with the results of his first blood test. "He has no viral load," she said. We all cheered; all except for Maya, that is, who was expecting a share in the treats he'd get after frequent blood draws and injections like she had experienced. For now, XiXi needs nothing but yearly monitoring.

Come what may, he's ours and we're his. His hepatitis B is a small part of the picture. We adore our son.[41]

Stories like this one tell the real tale of the struggles of patients with chronic hepatitis B. Callers to the Hepatitis B Foundation will always hear a compassionate response with accurate information and understanding of the challenges patients and their families face on a daily basis.

---

41   B Informed, Hepatitis B Foundation Newsletter, No. 60, (Fall, 2011.)

Viral Load

The story above also highlights the need to continually monitor a patient's viral load. Monitoring the viral load helps to determine if the patient's treatment is working and the severity of the patient's illness. In September 2008, the U.S. Food and Drug Administration (FDA) approved a new, more sensitive test to determine HBV viral load in an effort to improve the HBV monitoring process.[42]

Because the goal of hepatitis B therapy is to treat until the virus is undetectable, it is critical for viral load monitoring tests to be able to quantify very low levels of virus. Similarly, it is important for the test to quantify very high levels of virus (higher than 100 million IU/mL), an indicator of the need for more or less aggressive treatment.[43] Left untreated, chronic hepatitis B infection can result in complications such as cirrhosis (scarring of the liver) and liver cancer.

Experts in liver disease have found that the amount of hepatitis B virus in the blood, known as the viral load, helps determine the likelihood of developing these complications. Higher viral loads are associated with an increased risk of developing cirrhosis and cancer of the liver, so keeping the viral load as low as possible can help reduce or prevent injury to the liver. If liver cancer starts and has not spread beyond the liver, liver transplantation may be an option for some patients.

Liver transplantation, or hepatic transplantation, is the replacement of a diseased liver with a healthy liver. Liver transplantation is potentially applicable to any acute or chronic condition resulting in irreversible liver dysfunction, provided that the recipient does not have other conditions that will preclude a successful transplant. Uncontrolled metastatic cancer outside the liver, active drug or alcohol abuse, and active septic infections are absolute contraindications.

---

42 Hepatitis B Viral Load Measurement Improves. Hepatitis Central http://www. hepatitis-central.com/mt/archives/2008/09/hepatitis_b_vir.html (January 21, 2012)

43 http://www.hepatitis-central.com/mt/archives/2008/09/hepatitis_b_vir.html

Kenneth Rothstein, MD, chair of the gastroenterology and hepatology departments at Drexel University College of Medicine and also an HBF medical and scientific advisor, reminds us that "Extensive preparation and a team effort is needed to make sure that patients can get through their liver transplant."[44]

Living donor liver transplantation (LDLT) has also emerged in recent decades as a surgical option for patients with end-stage liver disease, such as cirrhosis and/or hepatocellular carcinoma, often attributable long-term untreated hepatitis B infection. The concept of LDLT is based on (1) the remarkable regenerative capacities of the human liver and (2) the widespread shortage of cadaveric livers for patients awaiting transplant. In LDLT, a piece of healthy liver is surgically removed from a living person and transplanted into a recipient, immediately after the recipient's diseased liver has been entirely removed.[45]

Whether it is a liver transplant, new medicines, or new therapies, patients and families have options. Patients and their families search for physicians and pour through clinical trials to gain understanding. This underscores the need for patient advocacy and providing the most up-to date information for callers to the Foundation. Advocacy, therefore, is another essential role of the Foundation, underscoring the commitment and the challenge to find a cure for chronic hepatitis B.

Ideally, connecting patient advocacy with helpful legislation would also assist families and bring hepatitis B awareness into the realm of national importance. Can advocacy be actionable?

---

44   Email and phone correspondence with Dr. Kenneth Rothstein (October 30, 31, 2011)

45   Intaraprasong P, Intaraprasong P, Sobhonslidsuk A, Tongprasert S. Donor outcomes after living donor liver transplantation (LDLT). J Med Assoc Thai. 2010 Nov; 93(11):1340-3.

4

PATIENT ADVOCACY

There are only two ways to live your life. One is as though nothing is a miracle. The other is as though everything is a miracle.
— Albert Einstein

The morning was frigid in Doylestown, Pennsylvania. At 6:00 a.m., it was still dark, save for the headlights of the black car idling in the driveway, the beams illuminating patches of slush scattered on the sides of the blacktop drive of the Block residence. Joan Block drove to the train station to catch the regional Amtrak train to Washington, DC. She was scheduled to testify before the Institute of Medicine (IOM) committee that was formed to study the prevention and control of viral hepatitis in the United States. Clinicians know that when the IOM provides a recommendation, it is generally understood as highly authoritative and is the "final word" on the topic.

The staff at the Hepatitis B Foundation has always been at the forefront of advocacy for patients and family members. Travel to the state capital of Harrisburg, Pennsylvania, to support legislation has also been a way to further the mission of the foundation and garner support. Also, the ability to travel to Washington, D.C. area and meet with legislators, write recommendations, and bring experts to the table has helped the foundation provide leadership and impart expertise to a wide audience willing to listen.

Advocacy in Action

One of the foundation's first significant efforts to advocate for patients, from a public health perspective, occurred at the Institute of Medicine (IOM) public hearing on Viral Hepatitis, on Thursday, December 4, 2008, in Washington, D.C. The IOM Committee heard presentations from Drs. John Ward, Dale Hu, and Cindy Weinbaum, who represented the Centers for Disease Control and Preventions Division of Viral Hepatitis (CDC/DVH), and also from Martha Saly and Chris Taylor, representing the National Viral Hepatitis Roundtable (NVHR).

Community presentations were given by Lorren Sandt from the Hepatitis C Caring Ambassadors Program and Joan Block from the Hepatitis B Foundation. Following the scheduled presentations, comments and testimony were accepted from the public. Arriving early, Joan was prepared to represent the nations hepatitis B community and provide urgent testimony about the needs of patients and families to the committee.

The IOM, funded by from the CDC/DVH and the NVHR, was charged with determining strategies to prevent new hepatitis B virus and hepatitis C virus (HCV) infections and reduce morbidity and mortality related to chronic HBV and HCV infections. The IOM also assesses the type and quality of data needed from state and local viral hepatitis surveillance systems to guide and evaluate prevention services. The final IOM report on viral hepatitis and liver cancer was published in January, 2010.

Joan Block testified before the IOM committee that the Hepatitis B Foundation represented 400 million people affected worldwide and that "there is an urgent need and unique opportunity to implement a zero tolerance policy toward hepatitis B." She discussed specific recommendations as to how we might be able to achieve this goal and continued by saying, "What the foundation means by 'zero tolerance' is that we should not tolerate *any* new infections and not allow any chronically infected individual to suffer from a lack of awareness or access to care.

We can strive for zero tolerance because we are now in the midst of a perfect convergence of opportunity: we have the tools, the means, and the leadership aligned to truly solve the problem of hepatitis B."[46] Joan detailed the many coordinated groups that have formed to create a "partnership advantage in our call to action." Lastly, she explained how "the biopharmaceutical industry is working together in the shared belief of the public health importance of hepatitis B."[47]

The light snow on the ground and chill in the Washington, D.C., air did not deter the committee, which also assessed current prevention and control activities and identified priorities for research, policy, and action, highlighting issues that warrant further investigations and opportunities for collaboration between private and public sectors.36

Patient advocacy, in a broad sense, is a very important part of the Hepatitis B Foundation. Through committee work, patient outreach, patient and family conferences, and all kinds of communications, the foundation is always looking for opportunities to solve problems. The Hepatitis B Foundation's medical professionals, scientific experts, and outreach staff regularly provide expert testimony about hepatitis B to state and federal legislators. The organization has successfully advocated for funds to support hepatitis B prevention, education and research programs.

Even before the IOM hearing in Washington, D.C., in December 2005, The first National Hepatitis B Act (S. 3558) was introduced in 2005 by Senator Rick Santorum (R-PA) and Senator Dianne Feinstein (D-CA) The act was designed to develop a national plan for prevention, control, and management of hepatitis B, including support for research and increased surveillance of chronic hepatitis B infection.[48] "This important legislation would be another great step forward in making sure that these Americans (with hepatitis B) will not 'B' forgotten," said Joan.

---

46   Hepatitis B: Time For Zero Tolerance. Testimony Submitted by Joan Block, RN, BSN http://www.hepb.org/pdf/hbv_iom_presentation.pdf (December 4, 2008)
47   Hepatitis B: Time For Zero Tolerance. (December 4, 2008)
48   Bucks County Herald, (Aug. 10, 2006)

FAST FACT:
An estimated one-third of those in the U.S. with chronic HBV acquire their infections as infants or as young children.

Also in that same year, during HBV Awareness Week, they provided national advocacy successes such as the HBV Bill, additional Congressional Briefings and became a charter member of the National Viral Hepatitis Roundtable. Dr. Block, his wife Joan and others from the Foundation are comfortable in Washington and have worked to build strategic solutions for patients with hepatitis B and look for ways to expand current research in an effort to find a cure. Through Congressional Briefings, testimony before key appropriations committees and meetings with key congressional leaders, Hepatitis B Foundation has always tried to provide the most accurate information to those who ask.

Joan's testimony has always been thoughtful and reflected years of experience listening to patients and families. Joan and other Hepatitis B Foundation staff speak openly and honestly to callers interested in learning more about the foundation, and they respond to all kinds of questions from patients and family members. In order to better represent advocacy efforts of the foundation and keep on top of the legislative efforts, the Hepatitis B Foundation signed an agreement in 2007 with Madison and Associates, a government relations firm, to monitor hepatitis B and liver cancer legistlative issues and coordinate meetings with key decision-makers in Washington, D.C.

Washington and the World

Three blocks from the capitol building in Washington, D.C., peppered with Japanese cherry blossom trees, is a refurbished historic brick bank building in a residential area that houses the Madison and Associates offices. J. Michael Hall and Alyson Lewis, partners at Madison and Associates, has worked closely with the Hepatitis B Foundation staff to analyze issues, and build strategic, thoughtful responses to hepatitis B issues. "We emphasize structuring client issues to help insure a high probability of success," commented Mr. Hall.

"Our job has been to help advance the mission of the Hepatitis B Foundation and its research institute, the Institute for Hepatitis and Virus Research (www.ihvr.org) that was established in 2004 and connect them with advocacy groups and leaders to educate them on the Hill.[49] That same year, Drexel University became the new academic partner of the Foundation. With this strong academic affiliation and new research institute, the Foundation had a lot of expertise to bring to the table and share with leadership in Washington, D.C.

"We [Madison Associates, LLC] have succeeded in doing so by assisting to increase Federal funding for the NIH and CDC with appropriations report language." "We are always connecting Hepatitis B Foundation leaders to educate friends on the Hill," said J. Michael Hall. "While we have helped open doors and identified opportunities, J. Michael Hall reflected, the Blocks have made the job both easy and fun due to their expertise and engaging presence..... Members and staff on the Hill like the Blocks.... their commitment and knowledge makes people want to help them."[50] "Our job, Michael continued, "has been to help the Hepatitis B Foundation and the Institute for the past 5 years."

"Tim and Joan have a lot of passion and naturally present the Hepatitis B Foundation as an organization that is cutting edge and growing.... The most memorable event was the time that we took Dr. Block and Anand Mehta to a meeting with an organization related to U.S. biodefense." The consultants at the meeting told Dr. Block about some research on glycovirs that they had read about in an effort to impress him. Dr. Block said, "That's me! I am Tim Block...that's my work!" The experts were very impressed and invited Dr. Block to submit a research proposal.

"While we have helped open doors and identified opportunities, the Blocks have made the job both easy and fun due to their expertise

---

49   J. Michael Hall and Alyson Lewis, Interview, (Aug. 10 & 11, 2012)
50   J. Michael Hall and Alyson Lewis, (Aug. 10 & 11, 2012)

and engaging presence.... Members and staff on the Hill like the Blocks.... Their commitment and knowledge make people want to help them," recalls Mr. Hall. "They are always a pleasure to work with and their commitment shows," comments Alyson Lewis.

Representatives from Madison and Associates routinely caucus with Dr. Block and Joan Block to include the opportunity to testify before the house and the senate Labor, Health and Human Services and Education Appropriations Subcommittee (Labor-HHS). Most large organizations, such as the CDC, the NIH, and the Public Health Service are assigned to one of the subcommittees. "One year Dr. Block testified and was highly respected by the audience," observed Mr. Hall. Another focus is to submit legislative report language related to chronic hepatitis B and liver cancer that is impactful and represents patients with the disease at the NIH and CDC.

In 2008, Dr. Block testified before the U.S. House of Representatives Committee on Appropriations Subcommittee on Labor, Health and Human Services, Education and Related Agencies. His aim was to highlight the urgent need to address the public health challenges of hepatitis B by strengthening programs at the Centers for Disease Control and Prevention and the National Institutes of Health. Dr. Block recognized Congressman Honda and others on the subcommittee as having been sympathetic and helpful to the cause.

Other Foundation historic advocacy successes include the First-ever National HBV Act introduced in Congress; first Congressional Briefing on HBV; and first National HBV Awareness Week in May called for by Congress. More recently, the Foundation was at the forefront in working with Congressional members to advocate for HR 3974, the Hepatitis and Liver Cancer Prevention and Control Act 2009, a comprehensive bill that addresses the needs of more than five million Americans living with chronic hepatitis B.

Collaboration was the key. Hepatitis B Foundation leadership continues to advocate for patients and families through testimony and report language related to chronic hepatitis B and liver cancer to support these patients and families in the U.S.

5

STEPPING OUT

Most people see what is and never what can be.
—Albert Einstein

From 1935 to 1955, Albert Einstein resided in Princeton, New Jersey, just off the shady green stretches of the prestigious Princeton University campus. The time-honored architecture calls to mind the universities of Oxford and Cambridge in the United Kingdom. Dignified and stately, the town conveys a mixture of traditional high Victorian Gothic–style buildings with stone archways, spired towers, and relief sculptures that frame the noble campus with traditions that harken back to the mid-1700s. The large entrance gates, by long-standing tradition, are only passed through by graduates.

Undergraduates are expected to use the gates on the sides of the large entrance until they have received their diplomas. To honor Princeton's status as a member of the Ivy League, each year the graduating class plants ivy on Nassau Hall, one of the oldest halls on campus. Nassau Hall was the meeting place for the Continental Congress during the summer of 1783, making Princeton the country's capital for four months.

The Princeton Workshop

It was only fitting that in 1995, the Hepatitis B Foundation brought together twenty-two of the nation's leading HBV experts to share current research findings close to the site where Einstein lectured to students. On a warm September day, the first Princeton Workshop began. Participants were invited to the workshop, designed to be a marathon think-tank about innovative HBV therapeutic strategies in a friendly atmosphere that would allow participants to share current research findings. The objective was also to exchange information that would have a real impact on the scientific community working on HBV. This unique event was organized by Dr. Block who brought leaders from academia, government and industry and was limited to twenty-five to thirty people, allowing for open, frank conversation.

Anand Mehta, D.Phil, associate professor at Drexel University College of Medicine, remarked, "The Princeton workshop is *the* HBV meeting.... Only the top people come, and the topics are focused on important issues regarding HBV clinical care." Dr. London described the workshops as "having a different focus each time; the earlier workshops concentrated on new drugs; then over time the workshops included topics such as biomarkers, liver cancer, cirrhosis, and other hepatitis-related interests. "In the beginning," Dr. London said, "nobody knew about the Hepatitis B Foundation, but within a few years of hosting the meeting, it became a big honor to be selected. It was private; it had a unique caché."

Discussions at the workshops included the need to study ducks and woodchucks, which have their own acute "animal" HBV infections that rapidly go away and leave little cell destruction. Other brainstorming sessions focused on T cells and their ability to modulate HBV gene expression in a non-cytotoxic fashion, causing rapid destruction of intracellular contents. The discussion was lively and so involved that breaks were cancelled and the debate continued through meals!

Photograph of participants at a recent Princeton Workshop held at the Center for Biotechnology in Bucks County. 2011 Princeton Workshop sponsored by the Hepatitis B Foundation: Front row (L to R): Timothy Block, Jake Liang, Hie-Won Hann, Bill Mason, Thomas London, 2nd row (L to R): Ju Tao Guo, Ira Jacobson, Harvey Alter, Diana Berard, Alison Evans, Anand Mehta, 3rd row (L to R): Haitao Guo, Stephan Menne, Nat Brown, Brian McMahon, Prakhash Bhuyan, Adrian DiBisceglie Back row (L to R): Simon Fletcher, Bud Tennant, Robert Gish, Raymond Schinazi, Jesse Summers, Stephen Locarnini. Photo taken outside HBF research center in Doylestown, PA (May 13, 2011)

Dr. Block's vision of the Princeton Workshop—sharing scientific progress, discussing roadblocks, and reviewing current research synergies in a small interactive workshop environment—turned out to be very meaningful. By the end of the day and a half discussion, the participants agreed that the workshop should be repeated the following year.

Dr. Blumberg attended the Princeton Workshop and enjoyed the discussions. Dr. Mehta remembers, "Honestly, when a Noble Prize winner shows interest in our work, it is priceless.... The confidence you get from that is hard to describe."[51]

---

51   Interview with Anand Mehta, Hepatitis B Foundation, (August 10, 2011)

The first Princeton Workshop was a great success, with new ideas exchanged, review of current progress in different areas of research, and lively debate of research challenges and where to advance next. The dialogue was all about the need to press on and share best practice findings. Group pictures were taken and collaborations between researchers strengthened.

With the success of the Princeton Workshop and its other endeavors, the Hepatitis B Foundation was definitely making its presence known in the world of hepatitis B research. The Princeton Workshop has become an annual event and the foundation has published the list of participants and the highlights of the discussions so others can benefit from the discussion. The conference remains small in number as Dr. Block originally envisioned, with active conversations about vaccine therapies, drug discovery, and combination therapy. The success of the Princeton Workshop captured the interest of the scientific community, and a new opportunity surfaced for the Hepatitis B Foundation to reach out to international researchers.

The International Meeting on the Molecular Biology of Hepatitis B Viruses provided a forum for researchers to share their discoveries, questions, and insights with their peers in a highly interactive environment. Dr. Block was chosen to coordinate the meeting in 2005 with Stephen Urban from Germany and in 2006, the Hepatitis B Foundation was asked to become the official coordinator of this prestigious international conference.[52]

Spreading Research Internationally

Although the original organizers could not have anticipated the success of this meeting, it has continued to attract scientists who are interested in all aspects of hepatitis B viruses, including biochemistry, molecular biology, traditional virology, immunology, pathogenesis, HBV-induced cancer, and the latest therapeutic advances against HBV and HDV.

---

52   International Meeting on Molecular Biology of Hepatitis B Viruses, http://www.hepb.org/hbvmeeting/meeting-history2011.html

Several unique aspects of the International Meeting on the Molecular Biology of Hepatitis B Viruses distinguish it from other scientific meetings. First and foremost, it encourages informal interaction among researchers at all stages of their careers. Second, it emphasizes work in progress by limiting presentations to unpublished data. International scientists studying diverse aspects of hepatitis B and D viruses, related animal model viruses, and hepatitis B–induced hepatocellular carcinoma present their findings. The meeting provides extensive opportunities for both junior and senior scientists to engage in joint discussions of the latest conceptual and technical advances regarding these viral pathogens and related diseases.

Finally, it provides an important training venue for graduate students and post-doctoral scholars, who have traditionally delivered the large majority of oral presentations. This emphasis on junior scientists is generally considered a major strength of the meeting; almost all poster presentations are given by trainees.

From the very first meeting, organized by Dr. Jesse Summers and Dr. Harold Varmus in 1985 and held at Cold Spring Harbor, New York, to the current Hepatitis B Foundation coordination of the meeting, scientists attend and share their knowledge. The Hepatitis B Foundation was also fulfilling one of its initial goals to become the go-to place for information on hepatitis B. The laboratory research, outreach, and office work continued to carry that vision.

Continuing to Step Out in a Fun Way

As the office hummed along, Joan Block and I had lunch in downtown Philadelphia on the second floor of the casual Parisian-style Caribu Café to brainstorm outreach efforts to increase the awareness of chronic hepatitis B. During the conversation, the idea of creating a mascot for the Hepatitis B Foundation surfaced. After the laughter subsided, phone calls to Disney proved helpful—a local expert in New Jersey was suggested by Disney's creative team. Joan Block and I met with him while our husbands and kids explored the Camden Aquarium. The mascot creator, who was formerly a part of the Disney team, was enthusiastic about the project, so we moved forward. Eventually, we settled on a large liver with a big engaging smile, white Disneyesque hands with four fingers, and typical mascot oversized shoes.

Although other potential names for the new mascot, which was partially funded by donations, were tossed about—"Bilirubin," for example—it was agreed that the perfect name was O'Liver (o-liver). O'Liver was a cheerful character who could confidently connect with audiences and assist with school-based vaccination efforts, legislative appearances, health-related parades, and other events. O'Liver mascot debuted in Harrisburg during Pennsylvania Hepatitis Awareness Month and also made an appearance at the first Hepatitis B Foundation Awards Gala. A star was born!

Kim Jungkind, Author and Ed Rendall pose for a picture with O'Liver during a parade in Philadelphia.

The Hepatitis B Foundation staff quickly realized the need for O'Liver marketing items, such as bandaids, pencils, and balloons. Elected officials always welcomed a greeting from this character during public events and official document signings, and O'Liver always traveled with an HBF volunteer "handler" to help navigate through the crowds and provide direction. Media photos of O'Liver and local and even national celebrities such as Mayor Ed Rendell, Pennsylvania senators Joseph Conti and Jim Greenwood, country singer Naomi Judd, and NBC10's Cherie Banks were displayed at the foundation's offices.

As the Hepatitis B Foundation continued to "Step Out," the board kept the focus on the research and scientific discoveries that were underscored at the Hepatitis B Foundation board retreat and ongoing quarterly meetings. The foundation continued to reach out to people locally and beyond. The *B Informed* newsletter was still released on a quarterly basis, each issue filled with news updates and information about local fund-raising events.

The outreach program also expanded with an office telephone line, e-mail help line, printed literature, and the launch of Drug Watch which proved to be very helpful for clinicians and the general public to keep on top of medication trials and emerging medications. Successful fund-raising efforts gave the organization the exciting opportunity to initiate and develop a more comprehensive educational outreach program to increase public awareness of hepatitis B.

6

EMERGING POLITICAL SUPPORT FOR HEPATITIS B VACCINATION

The important thing is not to stop questioning. Curiosity has its own reason for existing.
—Albert Einstein

The early morning was quiet in Doylestown. Driving past a beautiful residence and a classic Pennsylvania barn, through paved roads surrounded by wheat fields and geese flying over, State Representative James Greenwood left his home for a meeting with his constituents. Hepatitis B Foundation board members had reached out to Mr. Greenwood for his guidance and support for their research mission. Jim Greenwood met with the group, and arranged a meeting with Drs. Jay Hoofnagle and Leslye Johnson from the National Institutes of Health (NIH), in Washington, D.C., to discuss the research needs of viral hepatitis at the NIH Digestive Diseases Advisory Board. During this meeting, the founders and NIH leaders agreed that hepatitis research efforts should be better coordinated for the scientific community.

Dr. Block advocated for a center approach as the most advantageous route to achieving a cure for chronic hepatitis B, citing the successes of a past concentrated effort upon vaccine development in the 1970s. The need to conduct better disease surveillance was also suggested by Dr. Miriam Alter, Centers for Disease Control (CDC), who participated in the discussions. It was through surveillance that public health officials determined the failure of targeting only known risk populations for HBV vaccination, which ultimately led to the CDC's recommendation of universal infant vaccination against hepatitis B in 1991.

When Jim Greenwood became a Pennsylvania Senator, he introduced a bill, the Hepatitis B Prevention, Research, and Treatment Act (S.B. 1367), which reflected the need to improve vaccination, screening, treatment and research initiatives in the state. Although the bill did not initially pass, it was reintroduced in 1991 by state senator Richard Tilghman and passed. In 1992, the Pennsylvania Immunization Insurance Act (Act 35) also became effective, with resulted in more children getting vaccinated because the law required all health insurers to cover the cost of childhood immunizations for Pennsylvania residents, regardless of their insurance. Greenwood was elected to the U.S. Congress in 1992 and had become a federal advocate for the Hepatitis B Foundation.

On July 1, 1994, the New York State Legislature passed an amendment to New York's public health law, which includes mandatory immunization against hepatitis B for kindergarten entry. The ruling took effect immediately and applied to children born after June 1, 1993. This legislation was important for New York and showed the way for other legislative efforts to move forward and make an impact.

In 1996, Pennsylvania continued to lead the way with Governor Ridge signing into legislation Act 15, the Hepatitis B Prevention Act, which added the hepatitis B vaccine to the list of immunizations required for children to enter school. In addition, the law requires the Pennsylvania Department of Health to establish a statewide "catch-up hepatitis immunization program" for all students in the middle school as per the recommendation of the U.S. Department of Health and Human Services.

These legislative efforts helped professionals to ensure that all infants and children were protected against hepatitis B, which is a vaccine-preventable disease.

Vaccine Anniversary

2001 marked the 20th year anniversary of the world's first anti-cancer vaccine-the hepatitis vaccine...the hepatitis B vaccine. The Center for Disease Control and Prevention's Advisory Committee for Immunization Practices (ACIP) has protected millions of people to be immunized against hepatitis B virus through its published guidelines. Hepatitis B vaccination is the most effective measure to prevent infection and its longterm consequences that include cirrhosis, liver failure and liver cancer. Over the years with evidenced-based guidelines and supporting legislation, vaccine coverage against hepatitis B among children and adolescents has also increased substantially. Despite these successes, challenges remain.

To address these remaining challenges and accelerate progress toward elimination of hepatitis B in the United States, the ACIP has updated the hepatitis B immunization recommendations to expand beyond infants and children up to 18 years. Today, they now recommend that adults be vaccinated if they are at high risk for infection due to their occupation, lifestyle choices or other factors. A main focus is on universal infant vaccination beginning at birth, which provides a "safety net" for prevention of perinatal infection, preventing early childhood infections.

The Hepatitis B Foundation publishes a HBF Vaccine Watch at the hepb.org website and publishes information in the *B Informed* newsletter. Below is the most up to date table of vaccine information.

Table 1.

Hepatitis B Foundation Vaccine Watch

Hepatitis B Vaccines - Recommended for those at risk and patients with chronic HCV

NAME	VACCINE TYPE	COMPANY	STATUS
Engerix B	Recombinant HBV	GlaxoSmithKline Philadelphia, PA	Market, USA
Recombivax HB	Recombinant HBV	Merck, West Point, PA	Market, USA
Gen Hevac B	Recombinant HBV	Aventis Pasteur, Lyons, France	Market, Europe
Hepacare (formerly Hep-tagene)	HBV preS1, preS2	PowderJect, Oxford, U.K.	Market, Europe
Bio-HepB	HBV preS1, preS2	Biotech, Gen. Corp., Iselin, N.J.	Market, Israel
Heptavax Gene	Recombinant HBV	Berna Biotech, Switzerland	Market, Europe

Hepatitis A Vaccines -Recommended for those at risk and patients with chronic HBV and HCV

Havrix	Inactivated HAV	GlaxoSmithKline	Market, USA
VAQTA	Inactivated HAV	Merck	Market, USA
Avaxim	Inactivated HAV	Aventis Pasteur	Market, Europe

Combination Hepatitis Vaccines

TwinRix (Adult)	HBV and HAV	GlaxoSmithKline	Market, USA
Comvax (Pediatric)	HBV and HiB	Merck	Market, USA
Pediarix (Pediatric)	HBV, Polio, DTP	GlaxoSmithKline	Market, USA
Hexavac (Pediatric)	HBV, DTP, HiB, Polio	Aventis Pasteur	Market Europe

Hepatitis Vaccines in Development

Extra Strength Hep B (for poor responders)	Recombinant	GlaxoSmithKline (with Corixa)	Phase III
Hep B Vaccine	ISS-linked to HBsAg	Dynavax Technology, Berkley, CA	Phase I and II
Hep B DNA Vaccine Px	HBV DNA vaccine	PowderJect	Phase I

Reference: This Vaccine Watch chart, with more
recent updates, is available at hepb.org

With all of the strides made over the past twenty years, there have been some bumps in the road. Critics have suggested that the vaccine was linked to neurologic disorders (such as MS) and that the preservative, thimerosal, used in the vaccine could exceed safe mercury exposure levels for children. The thimerosal was removed from the vaccine. The safety of the vaccine was reaffirmed when the National Academy of Sciences' Institute of Medicine issued a report stating there was no evidence that the vaccine could lead to neurological damage.

Currently, there are a number of hepatitis B vaccines that are available from different companies and can be given separately or in combination with other vaccines. Hepatitis B vaccination is only part of the complete strategy to eliminate hepatitis B. Many people do not get screened for hepatitis B, are not vaccinated, and the virus can be spread by those affected who may not know they have the virus. Research, therefore, is needed to help those who have the disease and to halt the progress of the virus. Improved public health measures are needed to control and prevent the spread of the virus through vaccination, screening and surveillance.

With this in mind, Dr. Block's idea was to expand the Jefferson laboratory someday, and find more funding to expand his hepatitis research. But was it possible? Was there enough support to expand the vision?

7

RESEARCH MOVES NORTH

Knowledge is limited.
Imagination encircles the world.
—Albert Einstein

Hepatitis B Foundation was getting used to its headquarters on the beautiful, small, private, Delaware Valley College campus in Doylestown, Pennsylvania. The campus has sturdy stone buildings connected by sidewalks, squares of green grass, a patchwork of open farm fields, a student-run produce stand, and the charm of an agricultural college. "DVC was also pleased to have a handsome, three-story oblong stone research building that would be fully paid in ten years," Dr. Block remembers. DVC students could take advantage of a world-class research facility, including NIH-funded investigators and recognized staff on scientific editorial boards, members of the U.S. National Academy of Sciences, and even regular participation from a Nobel Prize winner in medicine—right on the DVC campus.

The Jefferson Center for Biomedical Research at Delaware Valley College campus and previous home of the Hepatitis B Foundation.

As the Hepatitis B Foundation settled into its new campus quarters, Dr. Block recalls that "we got off to a rapid success, discovering new candidate approaches to treating HBV, HCV, and other flaviviruses." Dr. Anand Mehta recalls, "Two new companies were created from our work: Synergy Pharmaceuticals and Nucleonics." As a board member at the time, I recall how pleased the Hepatitis B Foundation board was with the office space, conference room, classrooms, lab space, and adjacent convenient parking. Finally, an expanded laboratory space and outreach space to grow the foundation!

Outreach efforts were moving forward as well, with local grants to enhance information and increase the number of booklets, pamphlets, and activities organized by the Hepatitis B Foundation. The outreach expanded at the new location in three areas:

1. The www.hepb.org website

2. A focus on local education in different targeted settings

3. A unique road trip that helped the citizens of Pennsylvania understand more about hepatitis thanks to a grant from the Commonwealth of Pennsylvania.

The foundation initiated the Hepatitis Matters Coalition road trip that included O'Liver (the Hepatitis B Foundation mascot); Kenneth Rothstein, MD, chair of the division of gastroenterology and hepatology at Drexel University College of Medicine as a clinical expert on hepatitis B; Dr. Block, our scientific expert; a representative from the American Liver Foundation and I as the nurse educator. The group traveled in an SUV caravan to the six regions covered by the Pennsylvania Department of Health of Pennsylvania to provide a three-hour seminar about hepatitis B using a lecture and quiz-show format. The road trip was fun and reiterated for the participants how important it was to raise awareness about hepatitis B.

Dr. Rothstein recalls, "We traveled across rural Pennsylvania, and it was a great opportunity to reach out to first responders, fire fighters, and public health coordinators and present clinical information on hepatitis."

Calls to the foundation continued to increase, and staff was added to handle the calls and mail information, at no charge, to those requesting information. In 2000, a Temple University Master of Public Health (MPH) student, Chari Cohen, joined the staff to work on a project as a part of her graduate research requirement. Exciting opportunities surfaced to survey participants of the on-line Hepatitis B Information and Support (HBlist.org) listserv and assess the role of the Hepatitis B Foundation in an effort to understand the needs of patients and their families communicating through this virtual support group. Chari's analysis of the data showed a need for patients and families all over the United States to connect and communicate on a personal basis. So the idea of a patient and family conference was born!

The Hepatitis B Foundation, in collaboration with the Hepatitis B Information and Support List sponsored B Informed 2001: A Gathering of Friends, which was the first meeting specifically dedicated to those living with or affected by hepatitis B. "People wanted to know about treatment and disclosure of their illness and a forum to get support," said Chari. "As a direct result of the needs assessment, the first patient conference was organized for three days on the campus of DVC."

From as far away as Spokane, Chicago, and Dallas, forty people attended the first national patient conference to share their stories, listen to updates from medical and scientific experts, and make new friends in the emerging hepatitis B community. Lecture topics during the conference included treatment, liver transplantation, blood tests, and stories about what it's like to live with chronic hepatitis B.

"The first B Informed Patient Conference was fun but very busy," recalls Chari. "I was three months pregnant, and Molli and I became the housekeepers for the participants who stayed in the DVC campus dorm rooms and needed soap, sheets, pillows, and toiletries, which we delivered. We outfitted every room and changed the sheets—we were total maids for the weekend and also lectured as part of the conference."

Chari remembers that the conference, "was an overwhelming positive experience for families. It had never been done before—it was a groundbreaking idea." The families stayed up all night and enjoyed the opportunity to speak with other families.

Since then, every year, Hepatitis B Foundation has sponsored a B Informed Patient Conference in other locations, including Seattle, LA, San Francisco at the Asian Liver Center, and Philadelphia. The crowds have grown over the years to more than 200 participants. The conference, still the only national conference for those living with chronic hepatitis B, allows patients and their loved ones, parents of children, and health-care providers to gather and share their stories and information in a caring and supportive environment.

The foundation was so thrilled with Chari's efforts that they offered her the opportunity to join the staff as a part-time employee with the title of program coordinator. It was the first time the Hepatitis B Foundation started thinking about the need to infuse more public health expertise into its programs.

As the website expanded, the Wittes came up with the idea of starting a research fellowship to support young scientists who work for the Hepatitis B Foundation. The Bruce Witte Research Fellowship was established by the Wittes in 2000. The Fellowship, in honor of Paul Witte son Bruce, who died as a young boy of an unspecified autoimmune disease, provided the opportunity to have this forward thinking research support at the Hepatitis B Foundation.

The list of Wittee fellowship recipients can be found in Appendix 3.

Another development was the creation of the National Directory of Liver Specialists, which was unique because each practitioner's listing, included the available hepatitis B therapies provided. The staff used this list as a resource for patients who called asking for referrals to respected physicians across the country. Researchers also determined the national available hepatitis B virus research priorities. Hepatitis B Foundation laboratories at Delaware Valley College continued with the research priorities for almost eight years.

Creation of a "Biotechnology Incubator"

Dr. Block commented, "I knew, even before we arrived on the campus of Delaware Valley College in 1998, that in order to be successful we would need a 'critical' mass of scientists and professionals."[53]

"In terms of space needed for growing research needs and attracting a larger scientific community," Dr. Block continued, "I knew that eventually we needed to grow. We needed to be surrounded by a professional scientific community. As attractive an environment as was the Delaware Valley College campus, our research center was not a research institution."[54]

---

53   Personal interview and written summary correspondence, Dr. Timothy Block, (January 11, 2012)
54   Dr. Timothy Block, (January 11, 2012)

After a few years, plans were then set in motion to create a bigger research center. More researchers wanted to be a part of the team, but space was limited. At one point, Delaware Valley College provided makeshift temporary space for overflow of equipment.

Dr. Block proposed the creation of a Biotechnology Incubator. Other nonprofit research organizations with an interest in translational research would be invited to locate with or near the foundation. Start up companies, including those that resulted from the foundation's own work, such as Synergy and Nucleonics, and those with an interest in its work, would locate at the new center. The Hepatitis B Foundation and other partners would benefit from the increased intellectual mass, the greater opportunities, and, of course, the rent—the foundation would own the property.

The expectation was that the like-minded, interactive, nonprofit laboratories derived from different institutions but unified by their philosophy of development and translational research would benefit from each other's presence. In turn, the intellectual community would be enhanced and possible synergies for the development of their discoveries would be aided by the presence of local for-profit companies. This unusual collaboration would serve as a model for other such regional endeavors.

The Proposal

Dr. Block wrote a proposal called "A Biotechnology Center in Bucks County." "Although Hepatitis B Foundation had no lease obligations to Delaware Valley College, I wanted to be respectful and appreciative of the investment Drs. Feldstein and Gonnella had supported and not draft a plan that would take the Hepatitis B Foundation and me away." Dr. Feldstein was the Interim President of Delaware Valley College at the time. Another other idea was to find "adaptive or re-used facility space," such as an old factory close by.

The first idea considered was to stay on the Delaware Valley College campus. DVC would "contribute the land (approximately ten acres) and the Hepatitis B Foundation would raise money for the anticipated 120,000 square foot congruence center, lab, lodging, and office complex, through a public-private partnership. Hepatitis B Foundation and Delaware Valley College would both have ownership. Frank Gillespie, a developer from Bethlehem with considerable experience in science parks and adaptive reuse, agreed to be the developer.

Dr. Block said, "I also submitted a proposal to the State of Pennsylvania to create the center, with DVC as a participant in the proposal. Dr. Thomas Leamer, the president who succeeded Dr. Feldstein, shared my excitement." Governor Ridge had been a good friend to the Hepatitis B Foundation, and our state senator, Joe Conti, was also very committed to health research and biotechnology.

Unfortunately, Mr. Gillespie passed away. Dr. Block: "I approached Mr. Richard Lyons, another highly regarded developer from the area, to ask of his interest. He had developed several parcels of space that were extremely attractive. He agreed, and we began new negotiations with Delaware Valley College, eventually reaching an agreement on how the property could be developed on their campus.

Governor Ridge stepped down after the terrorist attacks of September 11, 2001, to join the Bush administration as the first Secretary of Homeland Security, and Lieutenant Governor Mark Schweiker took over as the Pennsylvania Governor. Schweiker is from Middleton, in Bucks County, and he took an interest in Dr. Block's grant entitled "PA Biotechnology Center." Joe Conti, who had by now gained considerable prominence in the state senate, was instrumental in guiding Dr. Block through the submission process. The Hepatitis B Foundation received a call from the governor's staff that Dr. Block's proposal was to be funded.

Dr. Block called Ken Lipton, the Delaware Valley College board chair, to clarify the submission. Dr. Feldstein and I each met with Governor Schweiker, who assured us that if Delaware Valley College and the Hepatitis B Foundation could come to an agreement, he would fund the proposal. A short time later I was asked by the governor to provide a full presentation," said Dr. Block, "with drawings."

Dr. Block continued by saying that "Delaware Valley College and the Hepatitis B Foundation agreed to share in the grant proceeds and the biotechnology center it would build, fifty-fifty." The Hepatitis B Foundation and Delaware Valley College received a check from Governor Schweiker to develop the Center. "I am so proud to know that research to cure this terrible disease is being done right here in my home town," Schweiker said. He continued by saying "Based on the Hepatitis B Foundation's impressive track record, I am confident they will make a huge global impact."

There were quite a few meetings over several months to discuss the considerations of the new center. At one point, the inclusion of a museum in the new center was even considered.

During the process, Dr. Block was able to meet with Delaware Valley College faculty about their ideas. Pennsylvania Senator Joe Conti, Rob Loughery and Jim Greenwood encouraged Dr. Block to consider adaptive reuses in the community. The new approach was to review local properties and see if there were buildings that might be easily converted into a new biotech incubator facility.

Eventually, the foundation considered buying the old DA Lewis building on Old Easton Road, in Buckingham, which was an old candle factory. The facility was made up of two buildings with a combined square footage of 115,000, but initial negotiations caused them to lose one of the buildings. The larger of the two available buildings, however, was eventually secured.

The final partnership agreement created a new, 501(c)3 (non-profit) corporation with the mission of building a biotech center and defining ownership of the ten-acre property and facility on Old Easton Road.

As the building was being gutted and re-constructed, Board members and interested supporters were given tours of the large inside space of the building. The author of this book looked at the plans that were stretched out on a large table in the cavernous building and thought I was seeing the outline of a Jacuzzi in the building plans! "That's great and so forward thinking—to include a Jacuzzi for the staff and researchers. All of us had a laugh knowing that what I saw was actually small unusually shaped closet!

When I spoke to the Steve Cohen, Architect for the Foundation, he also laughed. Steve was the pioneering architect for the new Foundation Center from the consultation phase, design, drafting, and planning process. He has also continued to be active in the evolving need to renovate expanding space for the current building and the new Feldstein Pavillion.

Molli Conti remembers that "the new location seemed to please everyone; the site was designated as a manufacturing zone and it met all of the state requirements. Molli was thrilled to work with Steve Cohen on the color palette for the Center. "The end results were very exciting with vibrant colors throughout the building that matched the energy of the exciting research that was being done" she said.[55] Other projects that Molli and Steve worked on were the re-numbering of the building labs and office space following the decision to relocate the entrance of the building to better accommodate the space.

The New Center

The Pennsylvania Biotechnology Center of Bucks County, legally named the Bucks County Biotechnology Center, was created to provide a home to the Hepatitis B Foundation and to: (1) enable the research mission of the foundation as a vehicle to accelerate the pace of research to find a cure, (2) recruit and train young scientists, (3) expand the outreach program, (4) and raise the foundation's visibility as a world-class resource.

According to Dr. Block, president of the new center, "The Hepatitis B Foundation started the center because we hope to accomplish much more, in partnership with biotech companies in this space, than would be possible alone. We can get more research

---

55   Personal interview and email correspondence with Molli Conti, November 14, 2011.

done by bringing together like-minded scientists—whether from commercial settings or nonprofit environments."

The Pennsylvania Biotechnology Center of Bucks County was developed using the $7.9 million state award from then-governor Mark Schweiker and additional funding secured by the foundation and the College. The 62,000 square foot center was the result of a unique partnership between the Hepatitis B Foundation and Delaware Valley College, funded in part by a grant from the Commonwealth of Pennsylvania. While the original project called for building on ten acres of the college campus, the opportunity to purchase and renovate the abandoned warehouse on Old Easton Road and reuse an existing structure eliminated the need for new construction and spared almost ten acres of open space.

In the usual flurry of excitement from the local press, including a helicopter landing with Governor Schweiker, ribbon-cutting, and numerous pictures, the center opened on a beautiful sunny day in 2006. The center features twenty state-of-the art laboratories and millions of dollars in sophisticated equipment that scientists use to seek cures for serious diseases. The center is a place of discovery, education, and job creation, with a shared vision of sustaining the vitality and beauty of Bucks County and the region. It seeks to advance biotechnology, maximize synergies between nonprofit scientists and their commercial colleagues, and launch new ideas and discoveries that will make a difference.[56]

Description: A drawing of the Pennsylvania
Biotechnology Center of Bucks County

---

56   Dr. Timothy Block, (January 11, 2012)

The center is not only home to the Hepatitis B Foundation but also to biotech startup companies and non profit research, companies such as Immunotope, which has developed a promising cancer vaccine technology, and Nucleonics, a Horsham, Pennsylvania, company that was developing gene-based vaccines. The Hepatitis B Foundation offices are also located in the center.

Dr. Block describes the center as "a world-class resource for education and research." The facility stimulates innovative research and serves as a valuable economic driver in the state. "It helps budding entrepreneurs," described Dr. Block, "some of whom are 'displaced' professional scientists from either the result of pharmaceutical downsizing, or academic leaders searching for a second beginning as entrepreneurs."

On opening day, Dr. Block was quoted as saying, "The Center means new jobs for people, a place that entrepreneurs can launch their ideas, a place that our research can grow, and a place that Delaware Valley College and Drexel can do teaching."[57]

Professor Wigdahl, chair of the department of microbiology from Drexel University, commented that "Tim has passion, a strong vision and drive for hepatitis B and I wanted to support him in the development of the Hepatitis B Foundation and his role as a full time faculty member in the Drexel University department of medicine." He continued to say the "Tim is an outstanding scientist and we are thrilled to support him."[58] "This is a way for the Hepatitis B Foundation to move the science forward and help expand our efforts to assist those affected with viral hepatitis worldwide," Dr. Block said. "From this building for discovery will come great things."

What makes the new center unique is that it will house and nurture biotechnology startup companies and nonprofit organizations under one roof, as well as promote regional economic development, education, and training. In addition to the Hepatitis B Foundation and its affiliate the Institute for Hepatitis and Virus Research, the center is home to the Drexel Institute for

57   KYW News radio, (January 2006)
58   Phone interview, Professor Brian Wigdahl, PhD (January 3, 2012)

Biotechnology and Virology Research and Ben Franklin Technology Partners of Southeastern Pennsylvania.

Dr. Block continued, "In our second year, we acquired the other building, and we now have control of the entire ten-acre campus, 115,000 square feet of space—strangely close to what we had imagined in the original plans almost ten years ago!"

Joel Rosen, the Hepatitis B Foundation chairman of the board, said, "To help jump-start the Pennsylvania Biotechnology Center, the state of Pennsylvania kicked in a $250,000 grant and labeled the center a Keystone Innovation Zone, making it eligible for tax breaks, loans, and grants to develop business."

Roseanne Rosenthal, president of Ben Franklin Technology Partners of Southeastern Pennsylvania, commented, "Tim has been preparing for this day for many years."[1] Dr. Thomas Leamer, president of Delaware Valley College at the time, said that "establishing the center will allow the school to strengthen its life sciences programs and better prepare students for careers in biotechnology. The center will play a crucial role in the education of tomorrow's researchers."

Dr. Block commented, "People can come here with just an idea and some credibility, and we'll let them in. I see this place as a dream machine ... I love this building.... This is my dream."[59]

"Companies here want to be close to the pharmaceutical companies, because that's who their partners are going to be," said Ms. Rosenthal.[60] "It's like a NASA launch project," Dr. Block described to the Associated Press.[61] At the ribbon-cutting ceremony, Dr. Block remarked, "In this renovated space will come great things. Work by top scientists will only accelerate the discovery of a cure for hepatitis B, liver cancer, and more."[62] "It's a grand day for the people of Bucks County and Pennsylvania,"

59   (Philadelphia Biz Journal, ,p. 31 vol. 25, No. 28. (Sept. 1-7, 2006)
60   The Intelligencer, At the Forefront" (April 30, 2006)
61   TimesLeader.com Associated Press, Buckingham, PA 2006)
62   Mcall.com online, by Ann Wlazelek of the Morning Call, 2006)

enthused Pennsylvania Senator Conti during the opening ceremonies.

Annual Gala for All

Celebrations continue on an annual basis! Each year, the Hepatitis B Foundation sponsors a crystal ball "Gala" to honor those who have helped the growing organization with a particular challenge, acknowledgement of leadership related to the foundation, or a significant piece of the hepatitis B research puzzle. Attendees include a distinguished group of scientists, public health professionals, friends, family members, local celebrities, and industry supporters. Each year, those receiving awards enjoy dancing and a wonderful dinner with community, national, and international leaders in the fight against hepatitis B. The gala attracted a small gathering of one hundred guests in 1997 and grew to more than three hundred guests in 2011!

Molli Conti, who was the Executive Director of Hepatitis B Foundation (2002-2007) remembers a most interesting gala that was held at the new Biotech Center. "Tim [Dr. Block] was super-excited to have the community see the new center, although it was a work in progress. We had cocktails in the lecture hall just inside the entrance, and the only place we could host the dinner for more than two hundred people was in the unfinished space within the building.

The tables were so close together we could hardly get by the chairs." Molli continued by recalling, "The dance floor was somehow squeezed into the space, and I think the honoree was standing on a table to accept the award. It was not my choice, but Tim was ecstatic to show off the new center and everyone agreed it was a celebration of HBF moving from a small non-profit to a national research center." The unusual location for the gala that year seemed to underscore the commitment to research and add to the unique atmosphere for all contributing to the foundation's success. The list of those awarded for scientific and community contributions over the years at the Hepatitis B Foundation galas are included in Appendix 2.

Molli Conti (wife of State Senator Joseph Conti) was instrumental in expertly representing the foundation. Senator Conti and State Representative Kathy Watson presented the state proclamation in Harrisburg and declared Hepatitis B Month in 2006. "There's very little public awareness about the dangers of hepatitis in Pennsylvania, and we need to change that," said Senator Conti. The foundation appreciated all of the supportive efforts of Mr. Conti and many other elected officials to understand hepatitis B and support the Hepatitis B Foundation.

8

NEW HOPE ON THE HORIZON

The only real valuable thing is intuition.
—Albert Einstein

Away from Philadelphia's hustle and bustle, in a quiet corner office of a historic medical building, is a hard-working physician leader and Director of the Liver Disease Prevention Center, Division of Gastroenterology and Hepatology. Dr. Hie-Won Hann, professor of medicine, Jefferson Medical College, describes the goal of treatment for patients with chronic hepatitis B as "the need to prevent hepatocellular carcinoma [liver cancer]."

"During the last decade, great strides have been made in the treatment of hepatitis B virus infections," comments Dr. Hann. The World Health Organization estimates that nearly 350 million people in the world have chronic hepatitis B infection.[63] Without proper intervention for HBV infection, these HBV carriers are at risk for developing cirrhosis and/or liver cancer, which could lead to premature death.

Hepatitis B Treatment

Although they usually do not eliminate the virus except in rare cases, the Federal Drug Administration has approved seven

---

63   The World Health Organization. Hepatitis B. http://www.who.int/csr/disease/hepatitis/whocdscsrlyo20022/en/index1.html

drugs for treating chronic hepatitis B virus. In order for a treatment to be considered a cure, the affected individual must have a loss of hepatitis B virus from their body *and* must develop protective antibodies against it. Despite the lack of a complete cure, these drugs have been shown to significantly decrease the risk of liver damage from HBV by slowing down or stopping the virus from reproducing. Aside from the interferons, all of the oral medications must be taken for at least a year or longer. The approved hepatitis B drugs include:

1. *Interferon Alpha*—approved in 1991for both children and adults.

2. *Pegylated Interferon*—approved in 2005 for adults.

3. *Lamivudine*—approved in 1998 for children and adults.

4. *Adefovir Dipivoxil*—approved in 2002 for adults.

5. *Entecavir*—approved in 2005 for adults.

6. *Telbivudine*—approved in 2006 for adults.

7. *Tenofovir*—approved in 2008 for adults.

When prescribed for one or more of the above HBV treatments, doctors test their patients' viral load for two reasons: to establish a baseline level of infection and (during treatment) to monitor the patient's response to therapy.

Medications such as Interferon and also direct antivirals such as Entecavir, Lamivudine and Adefovir have become an important aspect of treatment for chronic hepatitis B. These agents work to stop the hepatitis B virus from replicating in cells and reduce the amount of hepatitis B virus in the blood. These antiviral medications can stop the progression of liver disease and prevent liver cancer.

The Hepatitis B Foundation medical and scientific advisory board has resolved that "All patients should be tested for hepatitis B before receiving chemotherapy or immunosuppressive therapy. This recommendation comes in light of the growing body of evidence in peer-reviewed literature that reactivation of an inactive HBV infection can occur after initiation of chemotherapy or long-term immunosuppressive therapies, and that HBV reactivation can greatly complicate treatment and even cause fatal outcomes."[64]

Below is a list of approved hepatitis B drugs in the United States, which can be found on the foundation website under Drug Watch at www.hepb.org/drugwatch.

Table 2.

Approved Hepatitis B Drugs in the United States

- **Interferon Alpha** (Intron A) is given by injection several times a week for six months to a year, or sometimes longer. The drug can cause side effects such as flulike symptoms, depression, and headaches.

- **Pegylated Interferon** (Pegasys) is given by injection once a week usually for six months to a year. The drug can cause side effects such as flulike symptoms and depression.

- **Lamivudine** (Epivir-HBV, Zeffix, or Heptodin) is a pill that is taken once a day, with few side effects, for at least one year or longer.

- **Adefovir Dipivoxil** (Hepsera) is a pill taken once a day, with few side effects, for at least one year or longer. Pediatric clinical trials are in progress.

- **Entecavir** (Baraclude) is a pill taken once a day, with few side effects, for at least one year or longer. Pediatric clinical trials are in progress.

- **Telbivudine** (Tyzeka, Sebivo) is a pill taken once a day, with few side effects, for at least one year or longer.

- **Tenofovir** (Viread) is a pill taken once a day, with few side effects, for at least one year or longer.

---

64   B Informed. Hepatitis B Foundation newsletter, (September 2011)

Below are the current patient assistance programs available at the writing of this book. For the most up to date information, please refer to www.hepb.org.

Table 3.

Patient Assistance Programs

Hepatitis Type	Pharmaceutical Company	Contact Information
Epivir HBV (Lamivu-dine)	GlaxoSmithKline	GSK Commitment to Access 866-265-6491 866-518-4357
Hepsera (Adefovir)	Gilead	Gilead Advancing Access 800-226-2056
Entecavir (Baraclude)	Bristol-Myers Squibb	BMS Patient Assistance Foundation 800-736-0003
Tyzeka (Telbivu-dine)	Novartis	Novartis Patient Assistance Now 800-245-5356
Viread (Tenofovir)	Gilead	Gilead Advancing Access 800-226-2056
Intron A (Interfon-Alpha)	Schering	Schering-Plough Cares 800-656-9485
Pegasys (Pegylated Interferon)	Roche	Roche Patient Assistance Program 877-757-6243

Additional Resources

Free Medicine Foundation	573-996-3333
Merck Vaccine Assistance Program	800-293-3881
Partnership for Prescription Assistance	888-477-2669
Prescription Discount Program	800-506-3725
Rx Assist	800-444-4106
Together Rx Access	888-477-2669

Reference: Hepatitis B Foundation website, www.hepb.org

Over the past twenty years, keeping track of new treatments and emerging therapies was always a challenge, not only for patients and families but for researchers as well. The Hepatitis B Foundation introduced Drug Watch in 2001 as a regular column in the newsletter, and it is now on the www.hepb.org website so that those who are interested can monitor the progress of current treatment. This information has been a keystone for patients and families.

Table 4.

Hepatitis B Foundation Drug Watch: HBV Compounds in Development

Fall 2011

INTERFERONS: Mimic naturally occurring infection-fighting immune substances produced in the body

FAMILY/DRUG NAME	MECHA-NISM	COMPANY WEBSITE	STATUS, USA
Intron A (Interferon alfa-2b)	Immu-nomodulator	www.merck.com	FDA Approved 1991
Pegasys (PegInterferon alfa-2a)	Immu-nomodulator	www.gene.com	FDA Approved 2005

NUCLEOSIDE ANALOGUES: Interfere with the viral DNA polymerase enzyme used for hepatitis B virus reproduction

Epivir-HBV (Lamivudine)	Inhibits viral DNA polymerase	www.gsk.com	FDA Approved 1998
Hepsera (Adefovir Dipivoxil)	Inhibits viral DNA polymerase	www.gilead.com	FDA Approved 2002
Baraclude (Entecavir)	Inhibits viral DNA polymerase	www.bms.com	FDA Approved 2005
Tyzeka (Telbivudine)	Inhibits viral DNA polymerase	www.novartis.com	FDA Approved 2006
Viread (Tenofovir)	Inhibits viral DNA polymerase	www.gilead.com	FDA Approved 2008
Clevudine (Levovir)	Inhibits viral DNA polymerase	www.bukwang.co.kr	Approved in S. Korea
Emtricitabine (FTC)	Inhibits viral DNA polymerase	www.gilead.com	Phase III
MIV-210	Inhibits viral DNA polymerase	www.daewoong.com	Phase II
Amdoxovir (DAPD)	Inhibits viral DNA polymerase	www.RFSpharma.com	Phase II

NON-NUCLEOSIDE ANTI-VIRALS:

NOV-205 (Bam 205)	Small molecule	http://novelos.com	Approved in Russia
LB80380 (ANA380)	Inhibits viral RNA polymerase	www.lgls.com	Phase II
NEW! Myrcludex B	Entry inhibition	Pending	Phase 1A, Germany
HAP Compound (Bay 41-4109)	Inhibits viral nucleocapsid	www.aicuris.com	Phase I

| REP 9AC | HBsAg release inhibitor | www.replicor. com | Phase I |
| Alinia (Nitazoxanide) | Small molecule | www.romark. com | Pre-clinical HBV |

NON-INTERFERON IMMUNE ENHANCERS:

Boost T-cell infection-fighting immune cells and the body's natural interferon production

Zadaxin (Thymosin alpha-1)	Immune stimulator	www.sciclone. com	Orphan drug pproval in U.S. for liver ancer
CYT107 (Inter-leukin-7)	Immunomodulator	www.cytheris. com	Phase I/IIA
DV-601	Therapeutic vaccine	http:// dynavax.com	Phase I
HBV Core Antigen Vaccine	Therapeutic HBV vaccine	www.ebse.com	Phase I

POST-EXPOSURE AND/OR POST-LIVER TRANSPLANT TREATMENT:

HyperHEP S/D	HBV immune-globulin	www.talecris. com	FDA Approved 1977
Nabi-HB	HBV immune-globulin	www.biotestp-harma.com	FDA Approved 1999
Hepa Gam B	HBV immune-globulin	www.cangene. com	FDA Approved 2006

*Sincere thanks to **Timothy Block, Ph.D.** (Drexel U. College of Medicine, Philadelphia, PA), **Nat Brown, M.D.** (Presidio, San Francisco, CA), **Brent Korba, Ph.D.** (Georgetown U. Medical Center, Rockville, MD), and **Raymond Schinazi, Ph.D.** (Emory U. Medical School, Atlanta, GA), for their regular review of the HBF Drug Watch Update.* Source: www.hepb.org/drugwatch

* The chart above is the current information as this book goes to print. Check the www.hepb.org website for more up to date information and drug trial details.

Pivotal studies reemphasize the need for an active anti-HBV therapy for chronic hepatitis B patients with viral replication as the ultimate prevention and/or delay for the development of hepatocellular carcinoma. The other key concept that patients and families learn about is antiviral resistance. Development of resistance is associated with loss of initial response to the therapy, at which point the hepatitis B virus rebounds, the patient may be symptomatic and require additional laboratory testing.

Drug Discovery

The Hepatitis B Foundation also has a drug discovery program within an academic framework, for which Andy Cuconati, PhD, is a project leader and associate professor and adjunct associate professor at Drexel. "I think hepatitis B is an open book as far as developing new kinds of therapies," said Dr. Cuconati. "Our drug discovery efforts are designed to yield potential drugs with novel underlying mechanisms."[65] "People that are being treated with current hepatitis B drugs are still not able to mount an immune response to the infection so we try to identify compounds that look promising in cell culture," says Dr. Cuconati.

"The goal is to identify molecules to treat liver cancer using liver cancer cell lines that will eradicate liver cancer cells but not destroy normal liver cells. It has been observed that one of the original molecules (HBF-0079) is actually quite effective at inhibiting tumor growth of cancer in an animal model. Another molecule that may be an inhibitor of hepatitis B surface antigen in patients with chronic hepatitis B is called HBF 2-15."[66]

Other major earlier discoveries in which Dr. Block has been involved, in antiviral research are the development of a novel microorganism based assay for antiviral drugs, called "MOBA."[67] Dr. Block provided the scientific leadership in the important discovery on N-nonyl DNJ, a plant sugar that prevents secretion of HBV from liver cells and for the development of MOBA, an inno-

---

65   Phone interview with Andy Cuconati, (January 5, 2012)

66   Dr. Cuconati (January 5, 2012)

67   "In the Spotlight" *B Informed* Hepatitis B Foundation Newsletter, p. 8 (Winter, 2000)

vative microorganism based assay used for screening antiviral drugs.[68]

Dr. Block received the WW Smith award for MOBA, which is patented and currently used in development to search for many antiviral agents. He was also one of the discoverers of the phenomenon called "co-transformation" of mammalian cells, on which his thesis is based. Co-transformation has been used universally for introducing foreign DNA into cultured cells.

"Its rare to have a eureka moment in discovery, Dr. Cuconati says... it takes time to understand research findings...that is why the current collaborative research model that Dr. Block created is so valuable and unique."[69]

With this drug discovery program at the foundation, can the Hepatitis B Foundation researchers find a cure for chronic hepatitis B? Would the research efforts be recognized as breakthrough? Is it possible to have a Nobel prize–winning scientist working in the center who will discover the cure for chronic hepatitis B?

---

68   "In the Spotlight" (Winter, 2000)
69   Andy Cuconati, (January 5, 2012)

9

FIRM FOUNDATION

Setting an example is not the main mean of influencing another, it is the only means.
—Albert Einstein

There is one way to test a firm foundation—build on it and see if it holds the weight. The Institute for Hepatitis and Virus Research (IHVR), also known as the Pennsylvania Commonwealth Institute, was opened by the Hepatitis B Foundation in 2004 as the next step in the organization's evolution. As an independent nonprofit research entity, the Institute for Hepatitis and Virus Research will serve as the research arm of the Hepatitis B Foundation, fulfilling its mission to use discovery science to find new therapies and early detection markers for viral hepatitis and liver cancer. It is charged with seeking to find cures for viral hepatitis, nurturing biotechnology entrepreneurship, recruiting and training future researchers, and providing public health outreach programs in partnership with the Hepatitis B Foundation.

According to founding president and director Dr. Block, "The institute researchers formed powerful alliances among leading state and academic institutions." These alliances create greater focus in the pursuit of common "mission-oriented" themes and activities that will help make the IHVR a world-class research facility.

To assure maximum impact for its efforts, the Pennsylvania Biotechnology Center houses the original Hepatitis B Foundation laboratory (established in 1994 at Jefferson Medical School), the IHVR and the Drexel Institute for Biotechnology and Virology Research of Drexel University College of Medicine in adjacent laboratories. Together, these groups of scientists collaborate on a variety of fronts, developing and testing new drugs and therapies focused on viral hepatitis, liver cancer, and other viral diseases in state of the art laboratories.

This progressive approach, with innovative programs such as the ability to use a large eighty-thousand chemical compound library and screening program for anti-hepatitis drug discovery, will encourage the growth of more research advances. The ongoing work of the Hepatitis B Foundation has been a strong foundation through its outreach, advocacy, and political strength. The organization's efforts resulted in the National Institutes of Health recognizing and including HBV and HCV on its growing list of cancer-causing agents.

On January 31, 2005, the viruses were listed in the report as known human carcinogens, because studies in humans show that chronic hepatitis B and hepatitis C infections cause liver cancer. Hepatitis B increases the patients risk for developing liver cancer 20 to 40 percent, and 80 percent of all liver cancer worldwide is due to chronic hepatitis B infections.

When asked about the uniqueness of the Institute for Hepatitis and Virus Research's design and the importance of its work, Dr. Block stated, "This is a very unusual design for a nonprofit research institute, because we need to think outside of the box in order to create novel and effective therapies against hepatitis B.

IHVR is filling an important role that will result in better understanding, greater experience, and advanced technologies to help us combat hepatitis B, liver cancer, and other serious viral diseases." A new website was constructed specifically for the research institute and can be found at www.ihvr.org.

Finding a New Research Home

There were quite a number of meetings over a several months to discuss the considerations of the new center. At one point, even a museum was considered to be part of the new building!" During the process, Dr. Block was able to meet with Delaware Valley College faculty about their ideas. During the course of the meetings, Pennsylvania Senator Joe Conti, Rob Loughery and Jim Greenwood encouraged us to consider adaptive re-uses in the community. The new approach was to review local properties and see if there were buildings that would be easily converted into a new laboratory-instructional facility.

Eventually, we considered buying the old DA Lewis Print Distribution center, on Old Easton Road, in Buckingham, PA which was an established inactive candle factory. There were two buildings that added up to 115,000 square feet, but initial dithering caused them to loose one of the buildings. We secured the larger of the two available buildings. The final operating agreement created a new, 501c3 (non profit) corporation with the mission of carrying out the plans and defined ownership of a facility on old Easton road. While the original project called for building on 10 acres of the college campus, an opportunity to purchase and renovate an abandoned warehouse became available. By using an existing structure, the need for new construction was avoided and spared almost 10 acres of open space on the Delaware Valley campus.

The Hepatitis B Foundation and Delaware Valley College to each own 50% of the property and each appoint an equal number of members to its Board. Hepatitis B Foundation appointed the President (a Board position), and in deference to its leadership creating the Center, Delaware Valley College appointed the Vice President (also a Board position).

As the building was being gutted and re-constructed, Board members and interested supporters were given tours of the large inside space of the building. The author of this book looked at the plans that were stretched out on a large table in the cavernous building and thought I was seeing the outline of a Jacuzzi! "That's great and so forward thinking—to include a Jacuzzi for the staff and researchers. All of us had a laugh knowing that what I saw was actually small unusually shaped closet! When I spoke to the Steve Cohen, Architect for the Foundation, he also laughed. Steve was the pioneering architect for the new Foundation Center from the consultation phase, design, drafting, and planning process.

Molli Conti remembers that "the new location seemed to please everyone; the site was designated as a manufacturing zone and it met all of the state requirements. Molli was thrilled to work with Steve Cohen on the color palette for the Center. "The end results were very exciting with vibrant colors throughout the building that matched the energy of the exciting research that was being done" she said.[70] Other projects that Molli and Steve worked on were the re-numbering of the building labs and office space following the decision to relocate the entrance of the building to better accommodate the space.

The New Biotechnology Center

The Pennsylvania Biotechnology Center of Bucks County, legally named "The Bucks County Biotechnology Center", was created to provide a home to and to enable, the research mission of the Hepatitis B Foundation (Hepatitis B Foundation) as a vehicle to accelerate the pace of research to find a cure, recruit and train young scientists, expand the outreach bringing together like-minded scientists – whether from commercial settings or non-profit environments."

The newly opened Pennsylvania Biotechnology Center of Bucks County off Easton Road in Doylestown encompasses the $15 million state award from then Pennsylvania Governor Mark Schweiker. The 62,000 square foot center was the result of a unique partnership between Hepatitis B Foundation and Delaware Valley College, funding in part by a grant from the Commonwealth of Pennsylvania. The building was built on 10 acres of land and while the original project called for building on 10 acres of the college campus, an opportunity to purchase and renovate an abandoned warehouse became available. By using an exist-

---

70   Personal interview and email correspondence with Molli Conti, (November 14, 2011)

ing structure, the need for new construction was avoided and spared almost 10 acres of open space.

In the usual flurry of excitement from the local press, a helicopter with Gov. Schweiker landed with on the lawn of the Delaware Valley College. There was a ribbon cutting and numerous pictures, the center opened on a beautiful sunny day in 2006!

The center features 20 state-of-the art laboratories and millions of dollars of sophisticated equipment that scientists would use to find cures for serious diseases. The center is a place of discovery, education and job creation with a shared vision of sustaining the vitality and beauty of Bucks County and the region. It seeks to advance biotechnology, maximize synergies between non profit scientists and their commercial colleagues, and launch new ideas and discoveries that will make a difference.

Picture of the Pennsylvania Biotechnology Center of Bucks County.

The Center is not only home to the Hepatitis B Foundation, but also to biotech start up companies and non profit research.

Companies such as Immunotope which has developed a promising cancer vaccine technology, and Nucleonics, a Horsham, PA company that was developing gene-based vaccines. The Hepatitis B Foundation office headquarters is also located in the center which serves as an incubator for start up companies.

Dr. Block describes the center as "a world class resource for education and research." The facility stimulates innovative research and serves as a valuable economic driver in the state. "It would help budding entrepreneurs, described Dr. Block, 'some of whom are 'displaced' professional scientists from either the result of pharmaceutical downsizing, or academic leaders searching for a second beginning as entrepreneurs."

On the TV news later that day Dr. Block was quoted to say "So that means new jobs for people, a place that entrepreneurs can launch their ideas, a place that our research can grow and raise its visibility as a world-class resource."

According to Dr. Timothy M. Block, President of the Hepatitis B Foundation and the new Pennsylvania Biotechnology Center, "The Hepatitis B Foundation started the center because we hope to accomplish much more, in partnership with biotech companies in this space, than would be possible alone. We can get more research done that will help us by and a place that Delaware Valley College and Drexel University can do teaching."[71]

"This is a way for the Hepatitis B Foundation to move the science forward and help expand our efforts to assist those affected with viral hepatitis worldwide," Dr. Block said. "From this building for discovery, will come great things. "What makes the new center unique is that it will house and nurture biotechnology start-up companies and nonprofit organizations under one roof, as well as promote regional economic development, education and training. In addition to the Hepatitis B Foundation and its affiliate the Institute for Hepatitis and Virus Research, the center will be home to the Drexel Institute for Biotechnology and Virology Research, and Ben Franklin Technology Partners of Southeastern Pennsylvania.

71   Brad Segall, KYW Newsradio Bureau Chief, 2006

Dr. Block described the "In our second year, we acquired the other building, and now have control of the entire 10-acre campus, 115,000 ft2 of space- strangely close to what we had imagined in the original plans, almost 10 years ago!"

"We continue to be an essential and growing source of scientific innovation through the Pennsylvania Biotechnology Center," Joel Rosen, Esq., the Hepatitis B Foundation Chairman of the Board said, "and we thank all of our many friends who have contributed time, talent and donations to ensure our success!"

"It's like a NASA launch project," Dr. Block described to the Associated Press.[72] "In this renovated space will come great things" Dr. Block said at the morning ribbon cutting ceremony. "Work by top scientists, he added, will only accelerate the discovery of a cure for hepatitis B, liver cancer and more."[73] "It's a grand day for the people of Bucks County and Pennsylvania" remarked Pennsylvania Senator Conti.

Steve S. Cohen was the pioneering architect for the new Foundation Center from the consultation phase, design, drafting, and planning process. He has also continued to be active in the evolving need to renovate expanding space for the current building and the new Feldstein Pavilion. In November 2007, the second building was opened with more space for science discovery. The Pavilion of the existing building is named after Dr. Joshua Feldstein in recognition of his valuable contribution in forging the unique partnership between the Hepatitis B Foundation and Delaware Valley College.

Gala in the New Center

O'Liver mascot debuted in Harrisburg during Pennsylvania Hepatitis Awareness Month and also made an appearance at the first Hepatitis B Foundation Awards Gala. Each year, Hepatitis B Foundation sponsored a "Gala" and honored those who helped the growing organization with a particular challenge, acknowl-

---

72 TimesLeader.com Associated Press, Buckingham, PA 2006
73 Mcall.com online, by Ann Wlazelek of the Morning Call, 2006

edgement of leadership related to the foundation or a significant piece of the hepatitis B research puzzle.

Attendees include a distinguished group of scientists, public health professionals, friends, family members, local celebrities and industry supporters. Each year, those receiving awards are always humbled to receive recognition, enjoy dancing and a wonderful dinner with community, national and international leaders related to hepatitis B. The Galas attracted a small gathering of 100 guests in 1997 and grew to over 300 guests in 2011!

Molli Conti, Executive Director of Hepatitis B Foundation from 2002 to 2007, remembers a most interesting Gala that held at the new Biotech Center. "Tim [Dr. Block] was super excited to have the community see the new center, although it was a work in progress. We had cocktails in the lecture hall just inside the entrance and the only place we could host the dinner for 200+ was in the unfinished space within the building. The tables were so close together we could hardly get by the chairs."[74] People were dancing with hard hats on the concrete floor, surrounded by unpainted walls and on the table were linen draped tables and fine china!

Molli continued by recalling "The dance floor was somehow squeezed into the space and I think the honoree was standing on a table to accept the award. It was not my choice but Tim was ecstatic to show off the new center and everyone agreed it was a celebration of HBF moving from a small non-profit to a national research center." The unusual location for the Gala that year seemed to underscore the commitment to research and add to the unique atmosphere for all contributing to the Foundation's success. The list of those awarded for scientific and community contributions are included in Appendix 2.

Molli Conti, wife of Pennsylvania State Senator Joseph Conti, expertly representing the Foundation. Pennsylvania Senator Joe Conti and State Rep. Kathy Watson presented the state proclamation in Harrisburg and declared Hepatitis B Month in

---

74   Personal interviews and email correspondence with Molli Conti, November 11, 2011.

2006. "There's very little public awareness about the dangers of hepatitis in Pennsylvania, and we need to change that," said Joe Conti.[75] The Foundation appreciated all of the supportive efforts of Pennsylvania's elected officials and their time to understand more about hepatitis B.

Also in 2006, when the new Pennsylvania Biotech Center opened, Hepatitis B Foundation staff shared the Center's Economic Impact Study that showed the Center is responsible for 537 jobs and more than $164 million being injected into the local economy within the first 2 years of opening.[76]

As the increasing research focus continued, the need for more research space within a new facility continued. Dr. Blocks' idea was to expand the Jefferson lab some day, as he explained to the Foundation and its researchers, and dedicate even more space to hepatitis research.

75   Bucks County Herald, p. 16, June 1, 2006
76   Hepatitis B Foundation Press Release, Hepatitis B Foundation Opens New Institute for Hepatitis and Virus Research (February, 17, 2005)

10

STRATEGIC EXPANSION

The only reason for time is so that everything doesn't happen at once.
—Albert Einstein

RoseAnn Rosenthal, president of Philadelphia's Ben Franklin Technology Partners of Southeast Pennsylvania has this to say about Dr. Block: "He always amazes me that he comes up with new ideas.... He has really created a little hub up there, clustered around biotechnology.... Without his efforts, it wouldn't be happening."[77]

Concerned about diminishing funds and interest in hepatitis B research, and wanting to take a focused, team-oriented approach to finding cures for viral hepatitis and liver cancer, the Hepatitis B Foundation took a bold step forward in 2004, creating the Institute for Hepatitis and Virus Research (IHVR), also known as the Pennsylvania Commonwealth Institute. More recently, the Drexel Institute for Biotechnology and Virology Research (DIB-VIR) of Drexel University College of Medicine is now adjacent to the IHVR labs. Since DIBVIR is focused upon virus and liver diseases, this relationship essentially multiplies the numbers of scientists working on the global problem of hepatitis B and C, and liver cancer.

---

77 At the Forefront" The Intelligencer, (April 30, 2006)

Another way that the Hepatitis B Foundation expanded its academic credibility was through a historic agreement with the University of Oxford to create an ongoing research training program that serves both organizations by helping to develop the next generation of leading research scientists. Called the STEER program (steering post-secondary students into science and technology entrepreneurship and research career paths) and seen as a major step in bringing two well-known and respected institutions together in the battle against viral hepatitis, the program is designed specifically to attract and prepare college graduates for careers in biomedical research, public health, and biotechnology.

Through this program, the Hepatitis B Foundation selects individuals to begin training at its facilities who are then eligible to apply for enrollment in studies that may lead to a doctoral degree from Oxford.

"It's very important to train people in basic research, and this program provides the wherewithal to do it," says world-renowned scientist Raymond Dwek, FRS, head of the department of biochemistry at Oxford.[78] "Oxford has one of the largest biochemistry departments in the western world, and the Scripps Research Institute in La Jolla, California, is the only other location in the United States [not counting the PA Biotechnology Center] where we have this program set up.[79] Achieving this agreement has been a triumph, and it's an honor for both the university and the foundation." Professor Dwek has served on the Hepatitis B Foundation's scientific board for the past twelve years and has provided strong leadership for this exciting new program.

"Our goal is to steer young scientists into research careers so that after a year or two in our labs they will then be eligible to apply through the Hepatitis B Foundation for enrollment into scientific doctoral studies at the University of Oxford," says Dr. Block.

---

78   Hepatitis B Foundation Press. Peggy Farley "Release Hepatitis B Foundation Offers Unique Oxford Scholar Program" (March 22, 2005)
79   Hepatitis B Foundation Press. (March 22, 2005)

FAST FACT:
5,000 Americans die each year form the complications of chronic HBV.

The Hepatitis B Foundation and University of Oxford training program provides college graduates the opportunity to gain experience in a research laboratory or in a public health field—either in the Hepatitis B Foundation labs or in the adjacent Drexel Institute for Biotechnology and Virology Research (DIB-VIR) labs. The program is sponsored by a U.S. Department of Education training grant, awarded through the efforts of Senator Arlen Specter, who has been a national champion for funding biomedical research.

"It's geared for people who are interested in graduate work but aren't sure about making the commitment yet," says Pamela Norton, PhD, Drexel University Associate Professor and the program's academic director. "Participants work full time as research technicians, yet there is a mentoring component with expectations of working closely with professional scientists to achieve career goals."

This program and other research programs at the center value everyone's expertise and collaborative environment. All of these programs, however, need funding to find a cure for chronic hepatitis B. Over the years, the foundation has been involved with some creative fund-raising efforts.

11

HEARTFELT FUND-RAISING

Imagination is more important than Knowledge.
—Albert Einstein

The blur of trees, warm breezes, and miles of blacktop roads stretched in front of John Ellis as he turned his upsetting HBV diagnosis into an inspirational 1,100-mile bicycle ride from Florida to Pennsylvania in June of 2008. Sweat beaded on his face as he cycled through long straight roads, country towns, and windy streets to the final destination. John and his friend Jamal rode bicycles and shared the journey to complete the Believe in the Cure Cycling Tour, designed to raise disease awareness and funds for the Hepatitis B Foundation. John says, "Twelve million people live right here in the United States with hepatitis B. My name is John Ellis, and I am one of those people." John came up with the idea of a bicycle journey and contacted the Hepatitis B Foundation to discuss donating the proceeds to their efforts.

John's Story in His Own Words

Two weeks before my sixteenth birthday in June 2006, I was diagnosed with hepatitis B. Initially, I was incredibly scared and confused, simply because I didn't know what having hepatitis B would mean for me. I had even received the vaccine just years before in middle school, and I didn't think that something like this could happen to me.

To be honest, I knew nothing about hepatitis B. However, contracting hepatitis B forced me to come to grips with many difficulties that most teens never have to imagine. But I am not alone.

In the search for a way to improve my health, I came across a fifty-dollar bike at a local cycling store. I started riding my bike everywhere—to school, to work, and more. Inspired by the love of riding, I had an epiphany: I discovered "touring," which is essentially backpacking on a bicycle.

I knew right away that I wanted to ride for a cause. People need to become more aware about hepatitis B and how it affects us.

I contacted the Hepatitis B Foundation and after meeting them personally, I decided on a cycling tour from my hometown of Pensacola, Florida, to Philadelphia, Pennsylvania, the location of the Hepatitis B Foundation.

Oftentimes, when people get sick, the hardest part is believing that things are going to get better. With this 1,100-plus mile cycling trip, I hope to raise awareness about hepatitis B. But more personally, I hope to prove to myself that I can overcome the obstacles placed in my path.

So even though there is still no complete cure for hepatitis B, if I believe in my heart that things will work out for the best, then who's to say I'm not cured?

With all of your help, we can together help raise awareness to help find a cure for hepatitis B![80]

The tour was a success! During the final few miles, Dr. Blumberg joined John and Jamal on his own bicycle and crossed the finish line at Boathouse Row in Philadelphia with them. Hepa-

---

80   Celebrating the Power of One to Make a Difference! http://www.hepb.org/believe_in_the_cure/johns_story.htm (June 2-23, 2008)

titis B Foundation staff was awaiting their arrival with a victory party and hundreds of supporters on a clear, blue-skied day.

John and best friend Jamaal prepare for their 1100-plus mile journey to raise awareness about hepatitis B. Reference: http://www.hepb.org/believe_in_the_cure/johns_story.htm

Another kind of fund-raising effort was a private dinner and tour of a highly coveted collection of Ferraris. A wonderful outside dinner on a large deck surrounded by green fields punctuated by strategically placed statues and welcoming hosts created a lovely evening for Hepatitis B Foundation guests. The cars were very highly polished and in mint condition as guests strolled by and thrilled at the opportunity to view such a unique collection of valuable red Ferraris. The evening was topped off by Thomas Sweet's delicious cold ice cream and chocolates!

Running was Adrian Elkins passion. Adrian and his family started work on the 5K Answer to Cancer Race in June, 2003, racing against time.

Adrian's Gift

As the youngest of five children, Adrian grew up running, trying to catch up with his brothers and sisters, in high school, he discovered cross country running after he went to college.

During Adrian's sophomore year at Southern Oregon University, he was unexpectedly diagnosed with liver cancer due to chronic hepatitis B. This became the toughest race ever. Adrian courageously underwent months of surgery, chemotherapy, and related treatments to beat the cancer. During the last two months of his life, he devoted his time to organizing a race to raise awareness about liver cancer and hepatitis B.

Adrian began his life as an abandoned baby at a Calcutta orphanage. At three months, he was adopted by the Elkins family from McMinnville, Oregon. Although he appeared normal, Adrian and his parents always knew that he was a carrier of hepatitis B due to a blood transfusion he received as a premature infant in India. However, the never expected the disease to manifest itself as a rare type of liver cancer called hepatocellular carcinoma (HCC). It was devastating news. Still, there was hope. Adrian was an excellent candidate for surgery to remove the cancer, since it appeared to involve only the right lobe of his liver. He started his first round of chemotherapy while waiting for surgery.

Sadly, the operation brought more bad news. The cancer wasn't confined to the right lobe after all – the left lobe was affected too. Worse, the cancer has also spread to his lungs. Adrian battled his disease for ten months.

Adrian's final gift was the Answer to Cancer Race that will help generations of patients in their fight against liver cancer. He turned his passion into a legacy of caring that will endure.[81]

---

81   Adrian's Gift – The Answer to Cancer Race. Hepatitis B Foundation *B Informed*. Summer, 2004. p. 12.

Since its beginning, the Answer to Cancer Foundation has benefited organizations such as the Hepatitis B Foundation who are dedicated to hepatitis B and liver cancer. The annual race was started by Adrian Elkins, who died prematurely at age twenty from liver cancer due to chronic hepatitis B.

Another fund-raiser was the Nassau Swim Team. They raised funds for the Hepatitis B Foundation in honor of Andrew Wise, a former team member, coach, and lifeguard, who tragically died at the age of twenty-four from hepatitis B and liver cancer. Team members at the Nassau Swim Club in Princeton, NJ, an Andrew Wise Swim-A-Thon in tribute to his brief but shining life. Andrew died at age 24 years from liver cancer due to chronic hepatitis B.

Tee-ing up for a good cause is what family and friends did at the Wedgewood golf course in Coopersburg, PA to celebrate the 5th Annual Joseph Nagy Golf Tournament that benefits the foundation. Joe Nagy was an avid golfer who contracted hepatitis B through a blood transfusion and succumbed to complications from the disease. The tournament began as a tribute in memory of the well-loved man called "gramps."

On a lighter note, Concertmaster David Kim, the internationally acclaimed violinist, played "Motzart in Paris" for friends of the foundation in Philadelphia. He addressed the audience in Korean and English, expressing his commitment to the foundation mission and encouraging everyone to help support the foundations efforts.

I think Joan Block said it best: "Our heartfelt thanks go out to each of you for helping us to grow and achieve our dream of a research center dedicated to the cause and cure of hepatitis B. With ongoing support and generosity of our individual, corporate, and foundation contributors, we have made dramatic progress in this decade."[82]

---

82   B Informed, Hepatitis B Foundation Newsletter, pg. 2, Autumn 1997.

12

CHINESE CONNECTION: HAIMEN CITY

*The significant problems we face cannot be
solved at the same level of thinking were
at when we created them.*
—Albert Einstein

Soaring through the clouds and peering down to the large snake-like Yangtze River, the longest river in Asia, flowing from the glaciers on the Tibetan Plateau before emptying into the East China Sea at Shanghai, was the start of an insightful collaboration.[83] Researchers from Fox Chase Cancer Center have been flying into the Shanhai Pudong International Airport for years, traveling to a place 60 miles northwest of Shanghai and crossing Yangtze River to work with Chinese collaborators on a prospective cohort hepatitis study that was started in Haimen City.

In 1992, during the early time the Hepatitis B Foundation was forming, Dr. Thomas London MD, Director of the Liver Cancer Prevention Center at Fox Chase Cancer Center, Philadelphia, and Dr. Alison Evans, also working at the Fox Chase Cancer Center in collaboration with Dr. Fumin Shen from Shanghai Medical University and Drs. Gongchao Chen and Wenyao Lin, at

Haimen City Anti-epidemic Station.[84] They were embarking on what became an extensive NIH funded research study to determine the feasibility and benefit of screening for hepatocellular carcinoma (liver cancer). Haimen City was known to have high prevalence of chronic infection with HBV and hepatocellular carcinoma (HCC).[85]

The goal of the research was to identify the factors that led to higher risk of HCC. Chronic infection with HBV proved to be the major risk factor for HCC with viral load (quantity of HBV DNA in serum) the determining factor. Other risk factors were identified including family history of HCC, personal history of clinical hepatitis, and occupation (farmer), blood folate levels and selenium concentration in toenails were inversely associated with HCC risk.[86] The Haimen City cohort would continue for many years and yield findings that would contribute to the body of knowledge about chronic hepatitis B infection.

Little did they know at the time, their research became the largest cohort in the world for the study of liver cancer! Liver cancer, as seen clinically, is found in a subset of patients with hepatitis B virus. Research insights and findings contribute to the understanding of disease and also unlock clues to helping patients and more effectively manage patients with the disease.

Drs. London and Evans worked at the Fox Chase Cancer Center and became supporters of the Hepatitis B Foundation as it grew. Their research efforts to understand hepatitis B was important at both organizations. Dr. London became a member of the foundations board of directors and has been a key member with a clinical and scientific research background dedicated to hepatitis B. Dr. Evans joined the staff at the foundation, continuing her focus on the epidemiological findings and outcomes of the initial cohort hepatitis study and on-going studies.

84   Email correspondence from Dr. Chen, December 23, 2011
85   Hepatitis B Foundation Press Release, April 11, 2010, Hepatitis B Foundation Kicks off New International Campaign in Haimen City
86   Email and recorded correspondence, Dr. Thomas London, Summer, 2011

"Haimen County is one of the places in the world where hepatitis B virus (HBV) is a very visible disease, affecting a large proportion of the population and causing serious health effects for many. Working there [in China] gives renewed urgency to our work" reflects Dr. Evans, now an assistant professor in the department of epidemiology and biostatistics at Drexel University School of Public Health and also the director of public health research at the Hepatitis B Foundation.[87]

During an initial meeting in Haimen County years ago, Dr. Evans remembers that "when we walked down the center of the street in the early 1990s, we were followed by people in the town as we were westerners and looked different." At that time, Haimen County had low-rise buildings, travel in the small town was primarily by bicycle, and the town was surrounded by mostly agricultural areas.

By 1993, the study enrolled about nine-two thousand people and a data base was used to record the births dates and the cause of death of the study participants.[88] Dr. Shen, a Chinese genetic epidemiologist, was a professor at Shanghai Medical University (now Fujian University) School of Public Health. He led the public health students to collect the blood samples from patients for the research. Dr. London describes "The results of treating liver cancer, identified in people that are diagnosed after symptoms occur, are dismal; more than 50 percent of these high risk patients die within six months of diagnosis an only two to three percent of such patients survive five years."[89]

The one glimmer of hope is that surgical resection of small tumors, less than 3 cm in diameter, has led to prolonged survival. Tumors of this size rarely cause symptoms. They can be detected, however, by ultrasonography (US) or magnetic resonance imaging (MRI)."[90] With this in mind, it is clear that early detection is crucial.

---

87 Interview with Dr. Alison Evans, Hepatitis B Foundation offices (August10, 2011)
88 The work was supported by USPHS grants CA-40737, CA-90395 and CA-06927 from the NIH and by an appropriation from the Commonwealth of PA. Drs. Evans and Chen received support from the Bristol-Myers Squibb Pharmaceutical Research Institute.
89 Interview with Dr. London, Summer and Fall, 2011
90 Interview with Dr. London, Summer and Fall, 2011.

A total of 60,744 men living in Haimen City, who ranged in age from twenty five to sixty four, were screened as a part of the research for the presence of hepatitis B. Of the initial group of more than three thousand men, thirty-three were identified as suspect for liver cancer.[91] The patients identified as high risk for liver cancer were referred to medical centers for treatment options.

Many times, however, the identified men need to pay for their treatment with personal funds or reach out to family and relatives to help pay for expensive tests and treatment. Currently, many residents in China do not have universal health insurance to cover the additional testing, ongoing monitoring and treatment for hepatitis B.

Dr. Chen describes the challenge as a "huge lack of knowledge about the transmission and misconceptions about the disease. The Haimen City, once a county, is small and rural in nature. The Chinese government is trying to improve the economic situation by medical coverage of specialized tests and on-going monitoring but it is not optimal at this time."

By 2003, a hepatitis B 'viral load' study was initiated with the support from Bristol-Myers Squibb (BMS). Using blood samples to detect HBV "viral load," a clinician can easily monitor a patient's development of the disease. Monitoring the viral load became the "standard" for all clinical trials for HBV drugs by 2006.

Dr. Chen, director of China Programs at the Hepatitis B Foundation was born and raised in Haimen City, China. His research and public health outreach is dedicated to helping the population of Hiamen City understand more about hepatitis B. Through a thoughtful multiyear public health campaign, Dr. Chen "hopes to increase the knowledge of hepatitis B so that affected people can get ongoing support, testing, and treatment to decrease the chance of getting liver cancer."

---

91   Phone interview and follow up email correspondence, Dr. Chen, November 23, 2011, Dec., 2011.

His public health interventions include disseminating culturally appropriate educational materials including playing cards with hepatitis B messages, pamphlets, posters, billboards, drinking cups, etc. in an effort to raise public awareness about hepatitis B. The playing cards, a very popular pastime in China, display key messages about transmission, prevention, and treatment of hepatitis B on different cards.

These efforts were designed to be a unique and fun way to educate people as they play card games. In addition, trained health educators at Haimen City CDC presented educational seminars to all levels of healthcare providers, obstetricians, school nurses and health curricula teachers, community leaders in the village and governmental officials throughout the city.

By 2006, an important paper was published regarding the predictive value of HBV-DNA on liver disease progression. The study also showed that liver cancer was diagnosed in forty of these patients and subsequently 80% of these men died in one year's time.[92]

Dr. London explains that patients with these tumors could have been diagnosed and treated early. What was learned from the study was: risk prediction in chronic hepatitis B was difficult, liver cancer (hepatocellular carcinoma) risk varies widely and high viral load was a strong predictor of liver – specific mortality.[93] For physicians taking care of patients with hepatitis B, this information was very helpful and provided a way to effectively manage this chronic disease.

In 2008, the Foundation donated a PCR machine to Haimen City CDC. This is the first machine to detect HBV viral load in Haimen City. Then in 2010, the Foundation, in collaboration with Haimen City CDC, started a city-wide public health campaign that was funded by BMS Foundation.

---

92   Phone interview and email correspondence with Dr. Chen, December, 2011

93   Gang Chen, Wenyao Lin, et al. Past HBV Viral Load as Predictor of Mortality and Morbidity from HCC and Chronic Liver Disease in a Prospective Study. Am J Gastroenterology 2006;101:1797-1803

Other activities of the research include perinatal prevention where women receive complete hepatitis B screening and their case is monitored to make sure they get appropriate medicine and immunization shortly after the baby is delivered to prevent transmission. This research started on July, 2011 with currently 1,152 pregnant women screened over a period of three months found that chronic HBV infection status was confirmed in 56 (4.86%) of the pregnant women. [94]

This study and other studies are a part of a Hepatitis B Foundation strategy to provide best practices for effectively managing patients with hepatitis B. More specifically, the Foundation's overarching goal is to determine the high risk patient populations, screen patients, provide outreach education for patients and family, and then refer patients to their local physicians for on-going management.

Gateway to Care

Recently in 2011, Dr. Chen hosted leaders from the Bristol-Myers Squibb Foundation, including the president of BMS Foundation, Mr. John Damonti, during a visit to Haimen City. Two Hepatitis B Foundation leaders, Joan Block and Dr. London, participated in the visit and launch of the foundation's Gateway to Care Program.[95] This multi-year citywide initiative will raise public awareness about hepatitis B, educate hundreds of healthcare providers to improve screening and management and provide information to those living with this serious liver disease.

The BMS Foundation Grant provided a charitable grant of $400,000 to support an outreach and educational effort among the 1 million residents. The grant will bring together key Foundation staff, the Haimen City Health Bureau, Haimen City Centers for Disease Control and the Haimen City Hospital for Women and Children as partners to promote hepatitis disease prevention, outreach and treatment.

---

94   Phone interview and email correspondence with Dr. Chen, December, 2011
95   Hepatitis B Foundation Press Release, Hepatitis B Foundation Kicks off New International Campaign in Haimen City (April 11, 2010)

Today, Haimen County is called Hiamen "City", meaning "a gate of sea" in the Jiangsu Province of the People's Republic of China, Haimen is a prosperous and clean city of a million residents with a pleasant warm climate. It is endowed with educational and cultural institutions, including a new bridge to make traveling to the city easier from the Shanghai Airport. The city continues to grow with more collaborative research support and education about hepatitis B that will directly benefit patients and families through its comprehensive education and outreach efforts.

The launch of the Gateway to Care public health campaign in Haimen City, China on April 8, 2011 with city leaders. This new international program is made possible by a $400,000 educational grant from the Bristol-Myers Squibb Foundation.

The Foundation, through web site resources, all kinds of research and phone support, has been able to assist thousands of patients and their families. International collaboration through research, conferences, and sharing information has been a focus with many countries including China.

Hepatitis B is the world's most serious, common liver infection that chronically infects four hundred million people worldwide.[96] More than 100 million of those chronically infected live in China. In China alone, almost 500,000 people die each year from hepatitis b-related complications such as cirrhosis and liver cancer.

Philadelphia as a "Sister City"

The Foundation's ambitious "Gateway to Care" campaign was developed to address the growing severity of hepatitis B and liver cancer among Asian populations in U.S. and China. The Foundation selected two demonstration sites: Philadelphia, PA and Haimen City, China. Philadelphia has the 12th largest Asian population of any U.S. city, with 219,000 Asian residents. Philadelphia is the closest city to the Foundation and it is the "Sister City" and is also rich in Chinese culture.

As you enter the Chinese predominantly Asian American neighborhood located within the Center City district in Philadelphia, the brightly painted red and gold portal "Friendship Arch" can be seen as you drive through Qing Dynasty style entrance. Filled with business establishments, vegetable stands, and restaurants the Chinese community is always bustling with activity. The community has also been embraced by the Foundation and by local university hospitals to improve hepatitis B awareness.

Actually, local screenings for hepatitis B have involved twenty-six thousand Asians in Philadelphia and surrounding counties in the mid 1980s to early 1990s with the expertise of Dr. Hei-Wann Hann and other physicians from Thomas Jefferson University Hospital. Findings from a recent Philadelphia-based study indicated that 20-25% of HBV carriers would develop liver cancer. [97] Alison Evans, ScD, Dr. London and Dr. Hann were involved with this longstanding effort. Dr. Hann, Director, Liver Disease Prevention Center at Jefferson Medical College, is a well respected clinician who currently manages over three thousand

---

96 http://www.hepb.org/patients/general_information.htm Accessed December 26, 2011.

97 McGlynn, K. et al. Susceptibility to Hepatocellular Carcinoma is Associated with Genetic Variation in the Enzymatic Detoxification of aflatoxin B1 Proc. Natl. Acad. Sci. USA, Vol. 92; 2384-2387 March 1995. Genetics

patients and previously worked with Drs. London and Blumberg at the Fox Chase Cancer Center.

Drs. Hie-Won L. Hann and Mark Zern, both liver specialists at Jefferson and medical advisors at the foundation, supervised screenings in the Chinatown section of Philadelphia. "The incidence of hepatitis B is high in certain countries," said Dr. Zern, "such as Asia, Africa, and some Mediterranean countries." When you are doing a screening in an area with people from a country with a high incidence, you expect a higher rate." Dr. Hann said, "The 22 percent of hepatitis B patients found in Chinatown was expected because it is consistent with the 20 percent rate found in the general Chinese population." Dr. Hann concluded that the participants "are people who knew something was wrong and that's why they came [to the screening.]"

The Foundation staff and collaborators were also involved in the screening of patients for the presence of hepatitis B in Philadelphia as a part of the Gateway to Care Campaign that launched in Philadelphia in 2008.

These "sister cities" (Philadelphia and Haimen City) have been coordinating active approaches to educate and increase public awareness about hepatitis B utilizing popular public health tactics. Chari Cohen, Associate Director, Public Health at the Foundation, leads in community coalition building in Philadelphia.

FAST FACT:

Approximately one million people die every year from chronic hepatitis B and its related complications, making it the 10th leading cause of death worldwide.

The vision is to raise the profile of hepatitis B in Philadelphia through a number of grass roots fun educational activities.

More specifically, the plan is to reach the high risk populations, screen patients, provide outreach education for patients and family and then refer them to their local physicians for on-going treatment. The aim is to implement "best practices" for testing and linking patients for appropriate intervention provided through the healthcare system in Philadelphia and in Haimen City.

Chari Cohen launched the "Hep B Free Philadelphia" campaign in the summer of 2011. One event was a "flash mob" with participants emerging from the covered plaza area to an open plaza at City Hall. Short encouraging speeches were provided by public health officials and the participants were medical students from local medical schools, students, Hepatitis B Foundation leadership, etc. It was a fun event that created a press sensation on local news.

Another hepatitis B awareness tactic included Philadelphia Department of Health to join the efforts and provide free hepatitis B vaccinations to anyone that is screened through the outreach efforts and is found to be uninfected. Twenty city-wide community based hepatitis B screening programs across the US, including Philadelphia, tested individuals for hepatitis B and found an 8% prevalence infection rate.[98] Dr. Cynthia Jorgensen, Education and Training expert at the CDC's Division of Viral Hepatitis assisted to develop a partnership plan for Philadelphia that supported the Foundation's educational and outreach efforts.

Today and in the future, the Foundation's efforts in China and in Philadelphia have raised awareness, provided a successful public health model that other cities may adopt and save saved lives.

---

98   D. B. Rei, S.B. Lesesne, P. J. Leese et al. Community-based hepatitis B screening programs in the United States in 2008. Journal of Viral Hepatitis 2009, Blackwell Publishing p.1-9.

A DIFFERENT KIND OF LIBRARY

Only a life lived for others is a life worthwhile.
—Albert Einstein

In June 2011, a unique treasure trove arrived at the Hepatitis B Foundation's research institute. Merck & Co., Inc., donated its entire natural products library, along with an undisclosed grant, to the Institute of Hepatitis and Virus Research (IVHR), the research arm of the Hepatitis B Foundation. The library will give researchers, students, and entrepreneurs access to one of the world's largest collections of plant and biological specimens.

Dr. Block describes Merck's gift as "an extraordinary opportunity for the foundation's research institute, allowing us to make this priceless treasure available to scientists from all over the world." The thirty-year-old Merck Natural Products Library, which includes approximately one hundred thousand screen-ready extracts and represents approximately 60 percent of all known plant genera in the world, is considered to be one of the most diverse and best-curated collections of its kind in the world. "By making the natural products library available to the global scientific community, the foundation hopes to advance drug discovery programs worldwide," adds Joel Rosen, Esq., board chair of the Hepatitis B Foundation.

The phrase *natural products library* conjures up images of exotic plants living in a steamy hothouse with various bugs and bacteria. But the actual library consists of samples of chemicals that were extracted from plants and microorganisms collected from far-reaching places all over the globe.

Dr. Block adds, "We are able to create a screening center and provide access to this collection in a way that may lead to new discoveries and new companies in both Pennsylvania and New Jersey. Having [the Natural Products Discovery Institute] here will put us on the map ... We'll be the Marcus Shale of the drug discovery industry, except instead of drilling for gas, people will come here to drill for new drugs."[99]

The Merck natural products library is considered to be one of the most diverse and best curated natural products libraries in the world.

Natural products are compounds derived from living organisms. Approximately 45 percent of all Food and Drug Administration–approved products—including treatment for bacterial and fungal infections, high cholesterol, high blood pressure, cancer, and central nervous system disorders—were originally derived from natural products.[100]

Merck exited the natural products drug-discovery business in 2008, at a time when technology was changing how drug companies pursued new drug candidates. Although the process of delivery takes time, discoveries may still take place using the library. If a discovery happens, it will open up an opportunity to help patients with chronic hepatitis B.

"Merck's natural products library is also available to a wider scientific community with the hope it will assist others in the discovery of new therapies for patients," said Tony Ford Hutchinson, PhD, Merck senior vice president for vaccines research and –development. "Our agreement with IHVR will also keep this extensive thirty-year library collection alive and well."

---

99   Philadelphia Business Journal, Page 1, (Oct. 28th-Nov 3, 2011)
100 Philadelphia Business Journal, (Oct. 28th-Nov 3, 2011)

As part of the asset donation agreement with IHVR, Merck will continue to have access to the collection through the IHVR's open-use program, in which scientists and companies can apply to use the library to screen for potential new therapeutics. The collection will also be used for educational purposes to teach students about natural products and ensure students have access to this valuable resource.

"This collection will not only create a great benefit for our medical community, but also will promote greater investment in the state's budding biotechnology industry," said Pennsylvania state senator Chuck McIlhinney (R-Bucks), who helped position the IHVR for this gift. "I look forward to seeing the IHVR put this collection to good use as a valuable medical, educational, and economic development tool."

Dr. Wigdahl adds that "I am hopeful that it will lead to planting the seeds for future development of antiviral agents and/or immunomodulator compounds that might be involved in curbing hepatitis B disease and replication."

This 'different kind of library' has the potential to advance the goals of the Hepatitis B Foundation and the IVHR.

14

A LOOK INTO THE FUTURE

I never think of the future, it comes soon enough.
—Albert Einstein

The future of the Hepatitis B Foundation is promising! In so many ways, the foundation is poised to continue strengthening its growing leadership role in the biotechnology world. On the research front, the foundation has space for additional researchers to collaborate and continue to incubate start up companies. Even with the possibility of decreased NIH funds looming in the background, meaningful partnerships will continue with scientists from all over the world through the annual Princeton Workshop, international conferences and other research collaboration. The future may be the development of diagnostics and identification of biomarkers for the early detection of liver disease.

Advocacy efforts on behalf of patients and families will continue through connections at the patient/family conferences and expanding its social media program through Facebook and Twitter; the hepb.org website with numerous helpful resources and continued local, regional and international outreach. The foundation will reach out to other international countries, expand their website languages and continue to create the best in class educational resources.

Plans are in place to continue the outreach and educational efforts in Haimen City for the next three years and expand the website to include other languages, intensifying the reach of this already award-winning website! Currently, the website is listed on the first page of a Google search, and 50 percent of the visitors to the site are outside of the United States. The website also has about one million visitors per year, which is an incredibly great response of for a healthcare-related website.

The Hepatitis B Foundation staff plans to amplify advocacy/governmental relations to benefit those who struggle with hepatitis B through opportunities to support legislature that helps patients and families with hepatitis B.

The Institute for Hepatitis and Virus Research is moving forward with funding to improve the quality of life of those affected by viral hepatitis B and C through research and outreach. The goal is to become a world leader in research for HBV and HCV and to reinforce the institute's position as the nation's authority and portal of entry for information about hepatitis B.

As in many years, commitment was tested with continued stress on the region's economy, but the importance of the center as a source of scientific innovation and economic vigor continues just as significantly as in the past. Scientifically, the foundation's biomarkers of liver cancer were validated in a worldwide study, and its liver cancer drug was found to be effective in animals. With more than forty peer-reviewed publications over the past year and the acquisition of the Merck Natural Products Collection, the scientific discovery is in full motion.

"Pennsylvania is home to some amazing colleges and universities," said Governor Rendell, "and some equally impressive technology companies—firms that are breaking new ground in cutting-edge fields like biomedicine, nanotechnology, and computer science. By bringing these two sectors together, we are encouraging new startup companies that are creating jobs in some of the world's fastest growing fields. That's great news for the commonwealth's working men and women, and the state's economy."[101]

Discussions are underway for a stand-alone liver cancer website with the possible name of Liver Cancer Connect. Clearly, more information and research on liver cancer will be an important area for expansion in the future. This will allow treatment and research to have a laser focus. The commitment to helping patients with chronic hepatitis B never ends.

In 2006, the foundation employed about 60 people, and as of 2012 there are 240 people. Of the 240, only 7 full-time staff members are responsible for managing the day-to-day administrative work of the foundation. In the future, the number of researchers, educators, office staff and outreach specialists, I believe, will continue to grow.

The important role of disseminating information and guidelines such as the recent HBV Clinical Algorithms will continue to place the foundation in a leadership position. The B Informed newsletter is second to none with the most up-to-date information on new and emerging drugs, vaccines and research discovery.

A Unique Framework for the Future

The Foundation has clearly become the "go-to" organization for hepatitis B. I believe that the success of the foundation has not only been strong leadership but the framework built on collaboration and sharing best scientific practices. The model at the foundation has been key to its success. First, the Foundation has built an organization with a coordinated palette of scientific research. The core research programs that incorporate three overlapping divisions will continue to provide discoveries: early detection and disease, experimental therapeutics, and molecular pathogenesis.

---

101 Governor Rendell Launches Bucks County Innovation Zone: Announces $250,000 in Operational Funding in Yahoo Finance and in Office of the Governor, Commonwealth of Pennsylvania, Sept 7 Gov. Press Office.

An overlay of a large Drexel research footprint, the expertise to teach students and the active encouragement of incubator companies to join and collaborate under one roof to solve scientific research challenges related to hepatitis B—it's a unique approach to research. In addition, the researchers are interacting on a daily basis with the outreach professionals who are talking to the patients.

On the educational, advocacy and out reach front, the foundation clearly has represented the patients and families with hepatitis B in a passionate, caring way. Their efforts have been far reaching through high risk patient screening, public health awareness, testifying in Washington, D.C., thoughtful fundraising, website connectivity, national guideline development, scientific conference leadership, *B Informed* publications with a local and global reach.

The foundations goal, therefore, to improve the lives of those affected by hepatitis B through a comprehensive structure of research, patient advocacy and education has certainly been accomplished.

Along with these accomplishments and notoriety, the Foundation will need to continuously earn their leadership role in the scientific and medical community such as bringing together experts to create national recommendations about hepatitis B. The Journal of Family Practice, for example, has published the Foundation's consensus recommendations and algorithm for primary care providers regarding whom to screen for hepatitis B and when to pursue further evaluation and management.[102] More of these endeavors will need to take place to forward its mission. Current configuration and scientific curiosity will bring the discovery of a cure for chronic hepatitis B closer to fruition.

Dr. Blumberg saw this exceptional synergy, innovation, and collaboration and said the foundation "provides hope that the

---

102 McHugh JA, Cullison S, Apuzzio J, Block JM, Cohen C, Leong SL, London WT, McNellis RJ, Neubauer RL, Perrillo R, Squires R, Tarrant D, McMahon BJ. Chronic hepatitis B infection: a workshop consensus statement and algorithm. J Family Practice. 2011;60(9): E1-E8.

research here will provide a better therapy and that's a great comfort to a person to know that something can be done."[103]

Chari Cohen remarked that "The most memorable person was meeting Barry–he was a research pioneer –if there is something that you don't know, and you have a curiosity about it, that's enough to learn more." She continued, "It was the curiosity about the unknown [that Dr. Blumberg mentioned to me] and that's where the greatest discoveries might come from." As the foundation looks forward, it will continue to touch millions of lives and will continue to be a beacon of hope.

---

103 Hepatitis B Foundation 20th Anniversary Video, http://www.youtube.com/watch?v=LUk2mPFs_cI (March, 2011)

APPENDIX 1

TRIBUTES TO DR. BARUCH S. BLUMBERG
(JULY 28, 1925–APRIL 5, 2011)

The Hepatitis B Foundation's friend and champion, Dr. Baruch Blumberg died from a sudden heart attack on April 5, 2011 while attending the International Lunar Research Park Exploratory Workshop at Ames in California as a keynote speaker.[104] The World Health Organization has officially recognized July 28th as World Hepatitis Day which is the birth date of Dr. Blumberg. Below are tributes provided to the author in expression of appreciation to Dr. Blumberg.

Baruch S. Blumberg, MD, PhD

---

104 Mountain View Voice, Accessed: June 28, 2011, http://ht.ly/5mGaZ

I.

It is with great sadness and heavy hearts that we notify the hepatitis B community of the passing of Dr. Baruch S. Blumberg. Dr. Blumberg died suddenly on Tuesday, April 5, 2011. His discovery of the hepatitis B virus and invention of the first vaccine against hepatitis B, which resulted in the Nobel Prize for medicine in 1976, have been among the most important in the history of science and medicine. In addition to serving as senior advisor to the president of Fox Chase Cancer Center, Dr. Blumberg cofounded the Hepatitis B Foundation and served the foundation as a trustee distinguished scholar. His ongoing acts of support to the foundation will always be remembered and admired.

It has been one of the greatest professional privileges of my life to have known and to have worked with Dr. Blumberg. He was a wonderful mentor to me, and to all of us at the Hepatitis B Foundation who had the honor of knowing him. His curiosity and enormous intellect was always so motivational. He made it clear to all of us at the foundation how one life can do so much to benefit the world. Nothing will be the same without him, but so much has changed because of him. He will always be an example and inspiration for us all.

Dr. Timothy Block
President, Hepatitis B Foundation

II.

Baruch Blumberg's spirit left Earth in the ideal way. In Hebrew we use the words *mot beshikar*—death with a kiss and with no suffering.

Baruch was a remarkable man—a scholar, an administrator, a leader, and a visionary—and he was modest. He had a tremendous enthusiasm for exploring; he had his feet on the ground and his head in the stars. Both his public and private life related to the same person. The warmth and general affection that he showed to all his friends and colleagues and his kindness and sensitivity, together with his humor, made Baruch an exceptional man who was universally liked.

He combined intellectual and personal honesty, kindness, professional integrity, a great deal of common sense, imaginative skills, and goodness of heart to everyone. He was widely respected throughout the world and in many countries, such as Korea, which practically worshipped him, since his introduction of his vaccine for hepatitis B has saved millions of lives. He had a sense of fun and mischief and of honor that made being with him so exciting and so exhilarating. He loved interacting with United Therapeutics. He loved the company.

We shall celebrate and remember Baruch's achievements. We shall mourn the loss of a great friend and scientist. He leaves, as it says in the Talmud, the greatest crown of all: the crown of a good name. To paraphrase Yeats: our glory was, we had such a friend.

Professor Raymond Dwek, D. Phil., FRS,
The University of Oxford

III.

Right after completing my oncology fellowship at Harvard, I was looking for a position in the Philadelphia area. I had the good fortune to land in Barry's clinical research group at the Institute for Cancer Research, as it was called at the time. My original plan of trying out "research life" for just one year stretched to two years, then to an open-ended contract that eventually led to seventeen happy years of my life at Fox Chase under Barry's leadership.

The Liver Cancer Prevention Center, initiated by Tom London and joined by Barry and myself right here at Fox Chase Cancer Center, continued to grow at Jefferson Medical College since my move there in 1988. At this center, we conducted HBV screening of the Asian-American community in Delaware Valley, vaccinated uninfected individuals, and monitored those identified HBV carriers. When the need for vaccination became apparent,

Barry contacted Merck immediately. As an expression of respect for Barry—who developed the vaccine with Irv Millman—the company donated an astonishing twenty thousand vials of vaccine. Tom and I vaccinated thousands of children and adults. I have continued the HBV screening and vaccination program at Jefferson. People have been so grateful.

Like Bruce Smith, I had the privilege of having Barry as my boss, teacher, mentor, colleague, and finally a great friend. Just seven weeks before Barry's unexpected demise, he spoke at Jefferson's GI grand round and medical grand round. On both occasions, he inspired the students and faculty members—young and old—with his ever-passionate lectures featuring diagram after diagram illustrating one hypothesis after another; it was like the good old days at Fox Chase. It was amazing that his extraordinary enthusiasm had not diminished one bit over the years.

Four hundred million people in the world are infected with the HBV virus. Following Barry's discovery of the virus and his invention of the vaccine, antiviral drugs are being developed and have become readily available. With vaccination and antiviral treatment, we can prevent liver cirrhosis, and liver cancer and patients are surviving.

While I owe my career to Barry, millions of patients with hepatitis B will owe their lives to him forever.

Hie-Won L. Hann, MD,
Thomas Jefferson University Hospital

IV.

Baruch Samuel (Barry) Blumberg, the discoverer of the hepatitis B virus (HBV) and coinventor of the first vaccine to prevent infection with the virus, is the intellectual and spiritual inspiration for the Hepatitis B Foundation. Barry died suddenly on April 5, 2011.

He began his research career as a medical student at Columbia University when he took an elective in tropical medicine in Suriname. There he observed that filariasis was rampant but that only some of the many infected people showed signs of disease. Suriname had many different ethnic groups, and some of the diversity in responses was associated with ethnicity. That led him to wonder for the rest of his life why humans living in the same environment responded so differently to infectious agents.

After medical training in New York City, Barry went to Oxford University to study biochemistry with Alexander Ogston. In England, he was exposed to the history of scientific discovery and began formulating his scientific ideas. It was there that he found his scientific inspiration in the lives and work of the nineteenth century naturalists Charles Darwin and Alfred Russell Wallace.

Shortly after receiving his doctorate, Barry began a lifelong pattern of collecting blood samples everywhere he went, making observations about the people from whom they were drawn and often collecting samples of vegetation from their environments. His initial interest was in genetic polymorphisms in human blood, inherited variants of proteins or blood groups, which he believed were likely to be associated with human diseases. As a believer in the central importance of natural selection, he thought all such variants had to be important. Otherwise, they would not have persisted in human populations.

He had complete faith that his approach, identifying variants in human blood and then finding out what they meant, would be much more informative than starting with a disease and trying to identify its causes. It was this approach that first resulted in identifying an antigen on a lipoprotein in serum, and subsequently to a different antigen, the Australia antigen.

After coming to the Institute for Cancer Research (the precursor of the Fox Chase Cancer Center), Barry focused the efforts of his research group on learning the biological significance of Australia antigen. In a series of studies of diseases associated with the antigen, the group found that Australia antigen was closely associated with one form of viral hepatitis, later called hepatitis B. At the same time they observed that Australia antigen was a particle similar in appearance to a virus. That was enough information for Barry to begin to develop a unique vaccine, one that was prepared from antigenic particles in human blood. The patent for the vaccine was submitted in 1969 and granted in 1971. By 1975, before the vaccine had even been tried in humans, Barry predicted in print that the vaccine would not only prevent infection with the hepatitis B virus but that it would prevent liver cancer. Therefore, it would be the first cancer vaccine.

In 1976, Barry was awarded the Nobel Prize in physiology or medicine for "discoveries concerning new mechanisms for the origin and dissemination of infectious diseases." In 1989, he returned to Oxford to become the master of Balliol College, serving until 1994. Balliol was founded in 1263 and Barry was its first American master. In 1991, Timothy Block went to Oxford on a sabbatical to work in the laboratory of Raymond Dwek and also to discuss ideas with Barry about starting a foundation dedicated to discovering a cure for hepatitis B. These discussions led to the current structure of the HBF and the IHVR.

From 1999 to 2002, he was director of the NASA Astrobiology Institute. He continued his affiliation with NASA, and in 2008 he became a senior scientist at the NASA Lunar Science Institute. In 2005, he became the president of the American Philosophical Society, founded by Benjamin Franklin and the oldest learned society in the Americas.

Barry left a legacy of accomplishments that saved an enormous number of lives and prevented hundreds of millions of people from becoming ill with the hepatitis B virus. On every continent, his many friends and colleagues mourn his loss, but we at the Hepatitis B Foundation will profoundly miss Barry's insights into the biology, prevention, and cure of hepatitis B.

W. Thomas London, MD,
Emeritus Fox Chase Cancer Center

APPENDIX 2

HEPATITIS B FOUNDATION

LEADERSHIP AND SCIENTIFIC AWARD RECIPIENTS

Harvey J. Alter, MD
National Institutes of Health

R. Palmer Beasley, MD
University of Texas School of Public Health

Timothy Block, PhD
Drexel University College of Medicine

Baruch Blumberg, MD, PhD, Nobel laureate
Fox Chase Cancer Center

Nat Brown, MD
(formerly) Glaxo Wellcome

Alan Brownstein
(formerly) American Liver Foundation
Connelly Foundation

Joe Conti
(former) Pennsylvania State Senator

Raymond Dwek, DPhil, FRS
University of Oxford

Joshua Feldstein, PhD
Delaware Valley College

Bill and Melinda Gates Foundation

Joseph S. Gonnella, MD
Jefferson Medical College

James Greenwood
(former) Pennsylvania Congressman

Hie-Won L. Hann, MD
Jefferson Medical College

Jay Hoofnagle, MD
National Institutes of Health

Anna S. Lok, MD
University of Michigan Medical Center

W. Thomas London, MD
Fox Chase Cancer Center

William S. Mason, PhD.
Fox Chase Cancer Center

Brian McMahon, MD
Alaska Native Tribal Health Consortium

Harvey Rich, CPA
Blanche and Irving Laurie Foundation

Richard Rosenberger, Esq.
Souder Rosenberger, LLP

Raymond Schnazi, PhD
Emory University

Jesse Summers, PhD
University of New Mexico

Deborah Wexler, MD
Immunization Action Coalition

Paul and Janine Witte
Origenetics

Howard Koh, M.D.
Assistant Secretary, U.S. Department of Health and Human Services

APPENDIX 3

BRUCE WITTE DISTINGUISHED LECTURERS

The distinguished Bruce Witte Lecturer, a named lectureship established by Paul and Janine Witte, cofounders of the Hepatitis B Foundation, is selected each year and is a thought-leader in the science and medicine of hepatitis B.

2011—Adrian DiBisceglie, MD, St. Louis University Medical Center

2010—Jake Liang, MD, National Institutes of Health

2009—Stephen Locarnini, MD, Victorian Infectious Diseases Reference Laboratory

2008—Rafi Ahmed, MD, Emory University

2007—Stanley Lemon, MD, University of Texas Medical Branch

2006—John Taylor, PhD, Fox Chase Cancer Center

2005—Ben Yen, PhD, University California San Francisco

2004—Emmet Keeffe, MD, Stanford University Medical Center

2002—Frank Chisari, MD, Scripps Research Institute

2001—John Gerin, PhD, Georgetown University

2000—Raymond Schinazi, PhD, Emory University

1999—Professor Raymond Dwek, FRS, University of Oxford

APPENDIX 4

CHARLES SIGETY
FAMILY RESEARCH SCHOLARS

The Sigety Fellowship provides funding and recognition for an outstanding Delaware Valley College student to work and study in the laboratories of the Hepatitis B Foundation and the Institute for Hepatitis and Virus Research.

2005—Anne Marie Dougherty

2006—Peter Olivieri

2007—Nikki Barnes

2008—Shannon Koh & Mark Pinkerton

2009—Lisa Faber

2010—Patrick Cocchiarella & Cassandra Voegtlin

APPENDIX 5

HEPATITIS B SCIENTIFIC DISCOVERIES AND HEPATITIS B FOUNDATION TIMELINE

1965

- Discovery of the hepatitis B virus by Dr. Baruch Blumberg at Fox Chase Cancer Center

1969

- Discovery of the first vaccine against hepatitis B virus by Dr. Baruch Blumberg

1971

- Blood banks begin to screen donor blood for hepatitis B virus with test developed by Dr. Blumberg

1976

- Dr. Blumberg wins the Nobel Prize in medicine for his discovery of the hepatitis B virus

1981

- U.S. FDA approves first commercial plasma-based hepatitis B vaccine

1986

- Second-generation recombinant hepatitis B vaccines approved that are still currently used

1991

- Jan. 28 Hepatitis B Foundation incorporated as a 501(c)3 nonprofit organization in New Hope, Pennsylvania

- Interferon alpha approved as first drug for HBV

- CDC recommends universal infant HBV vaccination

1992

- *B Informed* newsletter launched

1994

- Hepatitis B Foundation research lab established at Jefferson Medical College, Philadelphia, Pennsylvania

1995

- First Princeton Workshop

1996

- Hepatitis B Foundation launches www.hepb.org

- The first HBF office is located in Jenkintown, PA

- Act 15 requires HBV vaccine be added to list of school entry requirements in Pennsylvania

1997

- Pennsylvania declares June Hepatitis Awareness Month

- CDC recommends middle school HBV immunization catch-up program

1998

- Hepatitis B Foundation moves into its own research center on DVC campus

- Lamividine approved as first oral drug for HBV

- First online HBV support listserv started

1999

- Hepatitis B Foundation receives Pennsylvania state grant to conduct statewide training

- Hepatitis B Foundation first testifies before Congress at vaccine hearings

2000

- Bruce Witte Fellowship established by Paul and Janine Witte

- College summer research internship program created

- National HBV research priorities established

2001

- First national *B Informed* patient conference

- First AASLD Practice Guidelines for HBV

2002

- Adefovir approved

- HBF receives $400,000 grant from NIH to expand its website www.hepb.org

-

2003

- NIH establishes Liver Disease Research Branch

- HBF receives $7.9 million grant from Pennsylvania to build its biotech center in partnership with Delaware Valley College

2004

- HBF forms new academic partnership with Drexel University

- HBF open new research Institute for Hepatitis Viral Research

2005

- Entecavir approved

- HBF national advocacy successes include hosting first Congressional briefing on hepatitis B

- Pegylated interferon approved

- NIH releases ten-year action plan for liver disease research (including Hepatitis B Foundation research priorities)

- First National Hepatitis B Act introduced in U.S. House of Representatives

2006

- Telbivudine approved

- First National Hepatitis B Act introduced in U.S. Senate

- HBF opens Pennsylvania Biotechnology Center

2007

- HBF asked to become the official organizer and sponsor of the International HBF meeting

2008

- Tenofovir approved

- First NIH consensus conference on management of HBV

2009

- Viral Hepatitis and Liver Cancer Prevention and Control Act introduced in U.S. House

- HBF and Drexel scientistsdiscover early detection biomarkers for liver cancer

2010

- First IOM report on hepatitis and liver cancer

- HBF published first national pediatric consensus recommendations for chronic HBV

- HBF launches Hep B Free Philadelphia

- Viral Hepatitis and Liver Cancer Prevention and Control Act introduced in U.S. Senate

2011

- Twentieth anniversary of Hepatitis B Foundation

- HBF launches first outreach program in Haimen City, China

- First HHS action plan for chronic hepatitis and liver cancer

- Addition of the Natural Products Library at HBF

Reference: *B Informed* Newsletter, Spring 2011

APPENDIX 6

HEPATITIS B OVERVIEW

What is hepatitis?

Hepatitis means "inflammation of the liver." Toxins, certain drugs, some diseases, heavy alcohol use, and bacterial and viral infections can all cause hepatitis. Hepatitis is also the name of a family of viral infections that affect the liver; the most common types are hepatitis A, hepatitis B, and hepatitis C.

What is the difference between hepatitis A, B, and C?

Hepatitis A, B, and C are infections caused by three different viruses. Although each can cause similar symptoms, they have different modes of transmission and can affect the liver differently. Hepatitis A appears only as an acute or newly occurring infection and does not become chronic. People with hepatitis A usually improve without treatment. Hepatitis B and C can also begin as acute infections, but in some people the virus remains in the body, resulting in chronic disease and long-term liver problems. There are vaccines to prevent hepatitis A and B, but there is not one for hepatitis C. If a person has had one type of viral hepatitis in the past, it is still possible to get the other types.

What is hepatitis B?

Hepatitis B is a contagious liver infection that ranges in severity from a mild illness lasting a few weeks to a serious, lifelong illness. It results from infection with the hepatitis B virus. Hepatitis B can be either acute or chronic.

Acute hepatitis B infection is a short-term illness that occurs within the first six months after someone is exposed to the hepatitis B virus. Acute infection can—but does not always — lead to chronic infection.

Chronic hepatitis B virus infection is a long-term illness that occurs when the hepatitis B virus remains in a person's body.

Statistics

How common is acute hepatitis B in the United States?

In 2007, there were an estimated 43,000 new hepatitis B virus infections in the United States. Many people don't know they are infected or may not have symptoms and therefore never seek the attention of medical or public health officials.

Has the number of people in the United States with acute hepatitis B been decreasing?

Yes, rates of acute hepatitis B in the United States have declined by approximately 82 percent since 1990. At that time, routine hepatitis B vaccination of children was implemented and has dramatically decreased the rates of the disease in the United States, particularly among children.

How common is chronic hepatitis B in the United States?

In the United States, an estimated 800,000 to 1.4 million persons have chronic hepatitis B virus infection.

How common is chronic hepatitis B outside the United States?

Globally, chronic hepatitis B affects approximately 350 million people and contributes to an estimated 620,000 deaths worldwide each year.

Transmission/Exposure

How likely is it that acute hepatitis B will become chronic?

The likelihood depends upon the age at which someone becomes infected. The younger a person is when infected with hepatitis B virus, the greater his or her chance of developing chronic hepatitis B. Approximately 90 percent of infected infants will develop chronic infection. The risk goes down as a child gets older. Approximately 25 to 50 percent of children infected between the ages of one and five years will develop chronic hepatitis. The risk drops to 6 to 10 percent when a person is infected over five years of age. Worldwide, the majority of people with chronic hepatitis B were infected at birth or during early childhood.

How is hepatitis B spread?

Hepatitis B is spread when blood, semen, or other body fluid infected with the hepatitis B virus enters the body of a person who is not infected. People can become infected with the virus during activities such as:

• Birth (spread from an infected mother to her baby during birth)
• Sex with an infected partner
• Sharing needles, syringes, or other drug-injection equipment
• Sharing items such as razors or toothbrushes with an infected person
• Direct contact with the blood or open sores of an infected person

• Exposure to blood from needle-sticks or other sharp instruments

Can a person spread hepatitis B and not know it?

Yes. Many people with chronic hepatitis B virus infection do not know they are infected since they do not feel or look sick. However, they can still spread the virus to others and are at risk of serious health problems themselves.

Can hepatitis B be spread through sex?

Yes. Among adults in the United States, hepatitis B is most com-
monly spread through sexual contact and accounts for nearly
two-thirds of acute hepatitis B cases. In fact, hepatitis B is fifty
to one hundred times more infectious than HIV and can be
passed through the exchange of bodily fluids, such as semen,
vaginal fluids, and blood.

Can hepatitis B be spread through food?

Unlike hepatitis A, hepatitis B is not spread routinely through
food or water.

What are ways hepatitis B is *not* spread?

Hepatitis B virus is not spread by sharing eating utensils, breast-
feeding, hugging, kissing, holding hands, coughing, or sneezing.

Who is at risk for hepatitis B?

Although anyone can get hepatitis B, some people are at greater
risk, such as those who:

- Have sex with an infected person
- Have multiple sex partners
- Have a sexually transmitted disease
- Are men who have sexual contact with other men
- Inject drugs or share needles, syringes, or other drug
  equipment
- Live with a person who has chronic hepatitis B
- Are infants born to infected mothers
- Are exposed to blood on the job
- Are hemodialysis patients
- Travel to countries with moderate to high rates of
  hepatitis B

If I think I have been exposed to the hepatitis B virus, what should I do?

If you are concerned that you might have been exposed to the hepatitis B virus, call your health professional or your health department. If a person who has been exposed to hepatitis B virus gets the hepatitis B vaccine and/or a shot called "HBIG" (hepatitis B immune globulin) within twenty-four hours, hepatitis B infection may be prevented.

How long does the hepatitis B virus survive outside the body?

Hepatitis B virus can survive outside the body at least seven days. During that time, the virus can still cause infection if it enters the body of a person who is not infected.

How should blood spills be cleaned from surfaces to make sure that hepatitis B virus is gone?

All blood spills—including those that have already dried—should be cleaned and disinfected with a mixture of bleach and water (one part household bleach to ten parts water). Gloves should always be used when cleaning up any blood spills. Even dried blood can present a risk to others.

If I had hepatitis B in the past, can I get it again?

No, once you recover from hepatitis B, you develop antibodies that protect you from the virus for life. An antibody is a substance found in the blood that the body produces in response to a virus. Antibodies protect the body from disease by attaching to the virus and destroying it. However, some people, especially those infected during early childhood, remain infected for life because they never clear the virus from their bodies.

Can I donate blood, organs, or semen if I have hepatitis B?

No, if you have ever tested positive for the hepatitis B virus, experts recommend that you not donate blood, organs, or semen because this can put the recipient at great risk for getting hepatitis.

Symptoms

Does acute hepatitis B cause symptoms?

Sometimes. Although a majority of adults develop symptoms from acute hepatitis B virus infection, many young children do not. Adults and children over the age of five years are more likely to have symptoms. Seventy percent of adults will develop symptoms from the infection.

What are the symptoms of acute hepatitis B?

Symptoms of acute hepatitis B, if they appear, can include the following:

- Fever
- Fatigue
- Loss of appetite
- Nausea
- Vomiting
- Abdominal pain
- Dark urine
- Clay-colored bowel movements
- Joint pain
- Jaundice (yellow color in the skin or the eyes)

How soon after exposure to hepatitis B will symptoms appear?

On average, symptoms appear ninety days (three months) after exposure, but they can appear any time between six weeks and six months after exposure.

How long do acute hepatitis B symptoms last?

Symptoms usually last a few weeks, but some people can be ill for as long as six months.

Can a person spread hepatitis B without having symptoms?

Yes. Many people with hepatitis B have no symptoms, but they can still spread the virus.

What are the symptoms of chronic hepatitis B?

Some people have ongoing symptoms similar to acute hepatitis B, but most individuals with chronic hepatitis B remain symptom free for as long as twenty or thirty years. About 15 to 25 percent of people with chronic hepatitis B develop serious liver conditions, such as cirrhosis (scarring of the liver) or liver cancer. Even as the liver becomes diseased, some people still do not have symptoms, although certain blood tests for liver function might begin to show some abnormalities.

How will I know if I have hepatitis B?

Talk to your health professional. Since many people with hepatitis B do not have symptoms, doctors diagnose the disease using one or more blood tests. These tests look for the presence of antibodies or antigens and can help determine whether you have acute or chronic infection; have recovered from infection; are immune to hepatitis B; could benefit from vaccination.

How serious is chronic hepatitis B?

Chronic hepatitis B is a serious disease that can result in long-term health problems, including liver damage, liver failure, liver cancer, or even death. Approximately 2,000 to 4,000 people die every year from hepatitis B–related liver disease.

Tests

What are antigens and antibodies?

An antigen is a substance on the surface of a virus that causes a person's immune system to recognize and respond to it. When the body is exposed to an antigen, the body views it as foreign material and takes steps to neutralize the antigen by producing antibodies. An antibody is a substance found in the blood that the body produces in response to a virus. Antibodies protect the body from disease by attaching to the virus and destroying it.

What are the common blood tests available to diagnose hepatitis B?

There are many different blood tests available to diagnose hepatitis B. They can be ordered as an individual test or as a series of tests. Ask your health professional to explain what he or she hopes to learn from the tests and when you will get the results. Below are some of the common tests and their meanings. But remember: only your doctor can interpret your individual test results.

Hepatitis B Surface Antigen (HBsAg) is a protein on the surface of the hepatitis B virus. It can be detected in the blood during acute or chronic hepatitis B virus infection. The body normally produces antibodies to HBsAg as part of the normal immune response to infection.

A positive test means:

• A person has an acute or chronic hepatitis B virus infection and can pass the virus to others.

A negative test means:

• A person does not have the hepatitis B virus in his or her blood.

Hepatitis B Surface Antibody (anti-HBs) is an antibody that is produced by the body in response to the hepatitis B surface antigen.

A positive test means:

- A person is protected or immune from getting the hepatitis B virus for one of two reasons:

  - he or she was successfully vaccinated against hepatitis B; or
  - he or she recovered from an acute infection (and can't get hepatitis B again).

Total Hepatitis B Core Antibody (anti-HBc) is an antibody that is produced by the body in response to a part of the hepatitis B virus called the core antigen. The meaning of this test often depends on the results of two other tests, anti-HBs and HBsAg.

A positive test means:

- A person is either currently infected with the hepatitis B virus or was infected in the past.

IgM Antibody to hepatitis B Core Antigen (IgM anti-HBc) is used to detect an acute infection.

A positive test means:

- A person was infected with hepatitis B virus within the last six months.

Hepatitis B "e" Antigen (HBeAg) is a protein found in the blood when the hepatitis B virus is present during an active hepatitis B virus infection.

A positive test means:

•   A person has high levels of virus in his or her blood and can easily spread the virus to others.

This test is also used to monitor the effectiveness of treatment for chronic hepatitis B.

Hepatitis B e Antibody (HBeAb or anti-HBe) is an antibody that is produced by the body in response to the hepatitis B "e" antigen.

A positive test means:

•   A person has chronic hepatitis B virus infection but is at lower risk of liver problems due to low levels of hepatitis B virus in his or her blood.

Hepatitis B Viral DNA refers to a test to detect the presence of hepatitis B virus DNA in a person's blood.

A positive test means:

•   The virus is multiplying in a person's body and he or she is highly contagious and can pass the virus to others. If a person has a chronic hepatitis B virus infection, the presence of viral DNA means that a person is possibly at increased risk for liver damage

This test is also used to monitor the effectiveness of drug therapy for chronic hepatitis B virus infection.

Treatment

How is acute hepatitis B treated?

There is no medication available to treat acute hepatitis B. During this short-term infection, doctors usually recommend rest, adequate nutrition, and fluids, although some people may need to be hospitalized.

How is chronic hepatitis B treated?

It depends. People with chronic hepatitis B virus infection should seek the care or consultation of a doctor with experience treating hepatitis B. This can include some internists or family medicine practitioners, as well as specialists such as infectious disease physicians, gastroenterologists, or hepatologists (liver specialists). People with chronic hepatitis B should be monitored regularly for signs of liver disease and evaluated for possible treatment. Several medications have been approved for hepatitis B treatment, and new drugs are in development. However, not every person with chronic hepatitis B needs to be on medication, and the drugs may cause side effects in some patients.

What can people with chronic hepatitis B do to take care of their liver?

People with chronic hepatitis B should be monitored regularly by a doctor experienced in caring for people with hepatitis B. They should avoid alcohol because it can cause additional liver damage. They also should check with a health professional before taking any prescription pills, supplements, or over-the-counter medications, as these can potentially damage the liver.

Prevention/Vaccination

Can hepatitis B be prevented?

Yes. The best way to prevent hepatitis B is by getting the hepatitis B vaccine. The hepatitis B vaccine is safe and effective and is usually given as three to four shots over a six-month period.

What is the hepatitis B vaccine series?

The hepatitis B vaccine series is a sequence of shots that stimulate a person's natural immune system to protect against HBV. After the vaccine is given, the body makes antibodies that protect a person against the virus. An antibody is a substance found in the blood that is produced in response to a virus invading the body. These antibodies are then stored in the body and will fight off the infection if a person is exposed to the hepatitis B virus in the future.

Who should get vaccinated against hepatitis B?

Hepatitis B vaccination is recommended for the following groups:

- All infants, starting with the first dose of hepatitis B vaccine at birth
- All children and adolescents younger than nineteen years of age who have not been vaccinated
- People whose sex partners have hepatitis B
- Sexually active persons who are not in a long-term, mutually monogamous relationship
- Persons seeking evaluation or treatment for a sexually transmitted disease
- Men who have sexual contact with other men
- People who share needles, syringes, or other drug-injection equipment
- People who have close household contact with someone infected with the hepatitis B virus
- Health care and public safety workers at risk for exposure to blood or blood-contaminated body fluids on the job
- People with end-stage renal disease, including predialysis, hemodialysis, peritoneal dialysis, and home dialysis patients
- Residents and staff of facilities for developmentally disabled persons
- Travelers to regions with moderate or high rates of hepatitis B
- People with chronic liver disease
- People with HIV infection
- Anyone who wishes to be protected from hepatitis B virus infection

In order to reach individuals at risk for hepatitis B, vaccination is also recommended for anyone in or seeking treatment at the following types of facilities:

- Sexually transmitted disease treatment facilities
- HIV testing and treatment facilities
- Facilities providing drug-abuse treatment and prevention services
- Health care settings targeting services to injection drug users
- Health care settings targeting services to men who have sex with men
- Chronic hemodialysis facilities and end-stage renal disease programs
- Correctional facilities
- Institutions and nonresidential day care facilities for developmentally disabled persons

When should a person get the hepatitis B vaccine series?

Children and Adolescents

- All children should get their first dose of hepatitis B vaccine at birth and complete the vaccine series by six to eighteen months of age.
- All children and adolescents younger than nineteen years of age who have not yet gotten the vaccine should also be vaccinated. Catch-up vaccination is recommended for children and adolescents who were never vaccinated or who did not get the entire vaccine series.

Adults

- Any adult who is at risk for hepatitis B virus infection or who wants to be vaccinated should talk to a health professional about getting the vaccine series.

For more information about hepatitis B and other vaccines, see http://www.cdc.gov/vaccines/recs/schedules/default.htm.

Is the hepatitis B vaccine recommended before international travel?

The risk for hepatitis B virus infection in international travelers is generally low, although people traveling to certain countries are at risk. Travelers to regions with moderate or high rates of hepatitis B should get the hepatitis B vaccine.

How is the hepatitis B vaccine series given?

The hepatitis B vaccine is usually given as a series of three or four shots over a six-month period.

Is the hepatitis B vaccine series effective?

Yes, the hepatitis B vaccine is very effective at preventing hepatitis B virus infection. After receiving all three doses, hepatitis B vaccine provides greater than 90 percent protection to infants, children, and adults immunized before being exposed to the virus.

Is the hepatitis B vaccine safe?

Yes, the hepatitis B vaccine is safe. Soreness at the injection site is the most common side effect reported. As with any medicine, there are very small risks that a serious problem could occur after getting the vaccine. However, the potential risks associated with hepatitis B are much greater than the risks the vaccine poses. Since the vaccine became available in 1982, more than 100 million people have received hepatitis B vaccine in the United States, and no serious side effects have been reported.

Is it harmful to have an extra dose of hepatitis B vaccine or to repeat the entire hepatitis B vaccine series?

No, getting extra doses of hepatitis B vaccine is not harmful.

What should be done if the hepatitis B vaccine series was not completed?

Talk to your health professional to resume the vaccine series as soon as possible. The series does not need to be restarted.

Who should *not* receive the hepatitis B vaccine?

The hepatitis B vaccine is not recommended for people who have had serious allergic reactions to a prior dose of hepatitis B vaccine or to any part of the vaccine. Also, it not recommended for anyone who is allergic to yeast, because yeast is used when making the vaccine. Tell your doctor if you have any severe allergies.

Are booster doses of hepatitis B vaccine necessary?

It depends. A booster dose of a vaccine is a dose that increases or extends the effectiveness of the vaccine. Booster doses of the hepatitis B vaccine are recommended only for hemodialysis patients and can be considered for other people with a weakened immune system. Booster doses are not recommended for persons with normal immune status who have been fully vaccinated.

Is there a vaccine that will protect me from both hepatitis A and hepatitis B?

Yes, there is a combination vaccine that protects people from both hepatitis A and hepatitis B. The combined hepatitis A and B vaccine is usually given as three separate doses over a six-month period.

Can I get the hepatitis B vaccine at the same time as other vaccines?

Yes. Getting two different vaccines at the same time has not been shown to be harmful.

Where can I get the hepatitis B vaccine?

Talk to your doctor or health professional or call your health department. Some clinics offer free or low-cost vaccines.

What is hepatitis B immune globulin (HBIG)?

Hepatitis B immune globulin is a substance made from human blood samples that contains antibodies against the hepatitis B virus. It is given as a shot and can provide short-term protection (approximately three months) against hepatitis B.

Pregnancy and Hepatitis B

Are pregnant women tested for hepatitis B?

Yes. When a pregnant woman comes in for prenatal care, she will be given a series of routine blood tests, including one that checks for the presence of hepatitis B virus infection. This test is important because women infected with this virus can pass hepatitis B to their babies during birth. But this can be prevented by giving the infant HBIG and the first hepatitis B vaccine at birth and then completing the series.

What if a pregnant woman has hepatitis B?

If a pregnant woman has hepatitis B, she can pass the infection to her baby during birth. But this can be prevented through a series of vaccinations and HBIG for her baby beginning at birth. Without vaccination, babies born to women with virus infection can develop chronic infection, which can lead to serious health problems.

How does a baby get hepatitis B?

A baby can get hepatitis B from an infected mother during childbirth.

Can a baby be protected from getting hepatitis B from his or her mother during birth?

Yes, almost all cases of Hepatitis B can be prevented if a baby born to an infected woman receives the necessary shots at the recommended times. The infant should receive a shot called hepatitis B immune globulin (HBIG) and the first dose of hepatitis B vaccine within twelve hours of birth. Two or three additional shots of vaccine are needed over the next one to fifteen months to help prevent hepatitis B. The timing and total number of shots will be influenced by several factors, including the type of vaccine and the baby's age and weight. In addition, experts recommend that the baby be tested after completion of the vaccine series to make sure he or she is protected from the disease. To best protect your baby, follow the advice of his or her doctor.

What happens if a baby gets hepatitis B?

Most newborns who become infected with hepatitis B virus do not have symptoms, but they have a 90 percent chance of developing chronic hepatitis B. This can eventually lead to serious health problems, including liver damage, liver cancer, and even death.

Do babies need the hepatitis B vaccine even if their mother does not have hepatitis B?

Yes. The hepatitis B vaccine is recommended for all infants. CDC recommends that the infant get the first shot before leaving the hospital.

Why is the hepatitis B vaccine recommended for all babies?

Hepatitis B vaccine is recommended for all babies so that they will be protected from a serious but preventable disease. Babies and young children are at much greater risk for developing a chronic infection if infected, but the vaccine can prevent this.

Reference: http://www.cdc.gov/hepatitis/b/bFAQ.htm#overview
Accessed: December 10, 2011

HBV Reactivation and Chemotherapy/Immunosuppressive Therapy

Recommendations	Comments
Possible effects of HBV reactivation	Can interrupt or halt chemotherapy. Can cause acute liver failure and death.
Why is screening of only "high-risk" groups inadequate?	Many people do not know they have HBV or may not provide full risk history. Clinicians often lack time to ask about all possible risk factors.
Do antiviral drugs reduce the effectiveness of chemotherapy/immunosuppressive therapies?	No, preventive antiviral drugs generally do not interfere with these therapies.
Does HBV testing delay start of these therapies?	No, test results can be back in twenty-four hours at most labs.

Reference: Hepatitis B Foundation Newsletter Sept. 2011

APPENDIX 7

ABOUT THE HEPATITIS B FOUNDATION, THE PENNSYLVANIA BIOTECHNOLOGY CENTER AND INSTITUTE FOR HEPATITIS AND VIRUS RESEARCH

About the Hepatitis B Foundation: The Hepatitis B Foundation, celebrating its 20th anniversary as the global authority dedicated to eliminating hepatitis B, is the only national nonprofit organization solely dedicated to finding a cure and improving the quality of life for those affected with hepatitis B worldwide through research, education and patient advocacy. It is headquartered in the Pennsylvania Biotechnology Center, which it created to accelerate its research mission. To learn more, visit www.hepb.org.

Hepatitis B Foundation

3805 Old Easton Road, Doylestown, PA 18902

Phone: 215.489.4900 • Fax: 215.489.4313

Email: info@hepb.org

About The Institute for Hepatitis and Virus Research (IHVR): The IHVR is an independent nonprofit research institute established in 2003 by the Hepatitis B Foundation to conduct discovery research and nurture translational biotechnology in an environment conducive to interaction, collaboration and focus. To learn more, visit www.ihvr.org.

About the Pennsylvania Biotechnology Center

The Pennsylvania Biotechnology Center is a nonprofit research organization created by the Hepatitis B Foundation in partnership with Delaware Valley College and dedicated to the creation of a world-class biotechnology center; to the promotion of regional economic development and job creation; and to the education and training of tomorrow's researchers. For more information, visit www.pabiotechbc.org.

APPENDIX 8

COMMENTARY ON
BIOMARKERS AND SELECTED
LIST OF PUBLISHED RESEARCH

Commentary on Biomarkers

The "war against cancer" has made huge strides forward in the past several decades, with an overall decrease in cancer rates. Long-term survival has dramatically improved as the result of earlier diagnosis and better treatment options. But for primary liver cancer – known as hepatocellular carcinoma (HCC) – progress has been limited and the prognosis remains grim. According to the National Cancer Institute, primary liver cancer has become the fastest growing cancer in the U.S., and the numbers are expected to rise even more in the next 20 years, due to surging cases of chronic hepatitis B and C infections. Liver cancer is currently the fifth most common cancer in the world, and ranks eighth among leading causes of cancer death for Americans.

If only, many experts say, there was a good way to find the cancer early enough to effectively treat it. "The ideal biomarker would be used in a blood test that could predict who has liver cancer or is at high risk," said Robert Gish, M.D., medical director of the Liver Transplant Program at California Pacific Medical Center in San Francisco. "It could also be used to determine both an individual's risk of recurrence or treatment response."

According to Paul Wagner, Ph.D., program director of the National Cancer Institute's Cancer Biomarkers Research Group, the difficulty in finding sound biomarkers is not peculiar to liver cancer. "It's easy to find a marker in the blood that is increased in half of the individuals with a certain cancer, but very difficult to find such a marker in all people with that cancer," he said. "A challenge is to find a marker that picks up most people with cancer but doesn't give a lot of false positives."

For decades, the most widely used biochemical blood test for liver cancer has been alphafetoprotein (AFP), which is a protein normally made by the immature liver cells in the fetus. Yet, this test is controversial because it is not highly sensitive or specific enough for liver cancer. "AFP is often elevated because of liver injury or regeneration, and doesn't necessarily indicate the presence or absence of liver cancer," Gish said, though it might be useful in detecting an increased risk.

Most Promising Biomarkers

"We want to able to say to patients who come in for screening, 'This biomarker is increased, and although we can't confirm anything by ultrasound, we think it's very likely that you have liver cancer,'" said Wagner. "That would be the next step if our markers look good in our validation study."

Looking Forward

Primary liver cancer, of which 80% is caused by chronic hepatitis B, is a growing public health problem and workshop participants agree that early detection is important for its effective management. Current methods, however, are very limited in usefulness or practicality. The high mortality associated with liver cancer - because by the time it is diagnosed, it is often unresponsive to treatment – makes the development of biomarkers an urgent unmet need. With the current five year survival rate of liver cancer less than 5%, early detection biomarkers would certainly save lives.

Selected List of Published Research

• Biomarkers of liver cancer enter advanced option development with Abbott Diagnostics (Block et al.)

- Biomarkers of liver fibrosis are good enough to advance to testing in 15,000 people with Johns Hopkins University in an NIH-funded study (Mehta et al.)
- Primabody, marker of infectious disease, validated in animal studies (Chen/Romano et al, in progress)
- Liver carcinoid drug successfully completes Phase II human studies (Shallubhal et al, in progress with Callisto, Inc.)
- Urine DNA test for colorectal cancer and polyps and liver cancer gets major award from the Coulter Foundation (Su et al.)
- Human gene products discovered that repress hepatitis B, C and other viruses (Guo et al, Journal of Virology 2009)
- New pathway of HBV DNA synthesis discovered (Guo et al, Journal of Virology 2009)
- New drug (alkylated porphoryn) discovered that inactivates HBV, HCV & HIV without harming mammalial cells (Guo et al submitted 2010)
- Glucovir drug enhances immune attack against hepatitis B in woodchucks (Norton et al, Hepatology 2010)
- Liver cancer drug to be tested in animals (Cuconati et al, in progress)
- New glucovir compound protects animals from Dengue disease (Chang et al, submitted 2010)
- Serum micro arrays can represent 8,000 different people on a single small "chip" (Chen et al, in progress)

The Hepatitis B Foundation research is conducted in partnership with Drexel University College of Medicine, whose labs are located at the Biotechnology Center.

Excerpt on the commentary of biomarkers was from Newsletter of the Hepatitis B Foundation, No. 48, p. 1-3 Spring 2007

Excerpt of selected list of published research from Hepatitis B Foundation Annual Report 2010.

APPENDIX 9

HBF HAPPENINGS: A RECENT TIMELINE OF EVENTS AND MILESTONES

2011

Watch the Hep B Free Philadelphia Video from First City-wide Hospital Testing Day for HBV!

Nov.—Watch the Hep B Free Philadelphia coalition in action, led by the Hepatitis B Foundation, as they worked together to screen more than two hundred at-risk patients at four of the major hospitals in Philadelphia during one very busy day!

High School Students Working in HBF Research Labs Reach Finals in Siemens Competition

Nov.—Two high schools students were finalists in the national Siemens Competition with a project that involved determining whether the HBV genome can be used to detect liver cancer at an early stage of the disease in order to improve survival rates. Michael Chen and Kevin Chen worked in the Hepatitis B Foundation's research labs on this graduate school level project for almost two years.

Dr. Roy Vagelos, Retired Chairman and CEO of Merck, Keynote Speaker at HBF Conference

Oct.—The Hepatitis B Foundation's research institute invited Dr. Roy Vagelos to give the keynote address at its annual regional biotech conference, where we announced the launch of our new

Natural Products Discovery Institute. Under Dr. Vagelos's leadership from 1985 to 1994, Merck & Co. had very successful drug discoveries from its natural products library, which was donated to the Hepatitis B Foundation in June 2011. Based on this generous donation from Merck, we have created the Natural Products Discovery Institute to make this priceless collection available to scientists around the world.

HBF Sponsors Citywide Hospital Testing Day for Hepatitis B in Philadelphia

Oct.—For the first time, a citywide hospital testing day for hepatitis B was organized in Philadelphia on Saturday October 22 by the HBF as part of the Hep B Free Philadelphia campaign to increase screening rates in the fight against hepatitis B and liver cancer. With twenty community-based organizations and one hundred-plus student volunteers from local colleges, medical schools, and graduate public health programs, more than two hundred at-risk patients were screened in one day.

HBF Consensus Recommendations for Primary Care Providers Published

Sept.—The Hepatitis B Foundation's evidence-based algorithm and consensus recommendations for primary care providers on whom to screen for hepatitis B and when to pursue further evaluation and management was published as an online exclusive by the *Journal of Family Practice* (Sept. 2011, 60[9]E1-8) .The HBF convened a workshop of expert physicians, nurse practitioners, and physician assistants to develop practical recommendations for use in busy primary care practices.

HBF Summer Internships Program Trains Seventeen Young Scientists

Aug.—The Hepatitis B Foundation completed training of seventeen high school and college students from its competitive and highly valued summer research internship program that is funded in part by an educational grant from Merck's West Point Charitable Contributions Committee.

Dr. Timothy Block Named to the 2011 PharmaVOICE 100 List

Dr. Timothy Block, cofounder, volunteer president, and director of the Hepatitis B Foundation, the Pennsylvania Biotechnology Center, and the Institute for Hepatitis and Virus Research, was named to the 2011 PharmaVOICE 100 list for serving as a positive contributor to the life sciences industry.

White House Commemorates World Hepatitis Day 2011

Aug. 1st marked the first official World Hepatitis Day established by the World Health Organization, which was recognized in a special event at the White House. HBF was invited to send representatives to the White House, including Chari Cohen, Dr. Timothy Block, and Joan Block. Dr. Ronald Valdiserri, HHS Deputy Assistant Secretary for Health, wrote about this historic day.

BMS Foundation Leaders Visit Hep B Foundation's Haimen City Project

Aug.—The Hepatitis B Foundation was pleased to host leaders from the Bristol-Myers Squibb (BMS) Foundation on a visit to our Gateway to Care Project in Haimen City, China. BMS Foundation provided a $400,000 charitable grant to support a city-wide outreach and education effort among the city's one million residents.

Univest Bank Makes Investment in Hepatitis B Foundation

June—The Hepatitis B Foundation is the proud recipient of a major charitable donation by Univest Bank and Trust Co. to support its important outreach and education programs.

Listen to Hepatitis B Foundation Live on WNPV 1440 AM.

June—Hepatitis B Foundation's Dr. Timothy Block and Peggy Farley were interviewed live by Darryl Berger of WNPV 1440AM as part of a special event sponsored by Univest.

"B A Hero" Flash Mob Takes Over City Hall

May—The HBF mobilized the Hep B Free Philadelphia coalition to host a public awareness event at Philadelphia's City Hall to launch the "B A Hero" theme that promotes testing and vaccination in the fight against hepatitis B during May National Hepatitis Awareness Month.

HBF Breaks Fundraising Record at Twentieth Anniversary Crystal Ball

May—The Hepatitis B Foundation raised a record-breaking $95,000 at its Twentieth Anniversary Crystal Ball with more than 250 VIP guests in attendance! Brad Segall, KYW Newsradio bureau chief, was honored with the HBF's inaugural Community Commitment Award and Univest National Bank was this year's premier sponsor.

Hepatitis B Foundation's Joan Block Honored by Penn Asian Senior Services (PASSi)

PASSi's 2011 Distinguished Health Leader Award was presented at their annual gala on May 20 to Joan Block, HBF executive director and cofounder, in recognition of her contributions to improving the health of Asian Americans.

Hepatitis B Foundation Celebrates Twentieth Anniversary

The HBF is proud to be celebrating in 2011 our twentieth anniversary as a leading global authority dedicated to eliminating hepatitis B worldwide through research, education, and patient advocacy! On January 28, 1991, the Hepatitis B Foundation was officially incorporated as a 501(c)3 nonprofit organization in New Hope, Pennsylvania.

2010

Hep B Free Philadelphia Hosts Screening and Vaccination Workshop

Dec.—Hep B Free Philadelphia hosted its first HBV Screening and Vaccination Training Workshop. It attracted more than forty participants from community-based organizations, public health professionals, and college and medical student volunteer groups. The goal was to train interested groups to help increase the number of HBV screening and vaccination efforts in the city, which is the mission of Hep B Free Philadelphia.

HBF Joins NVHR to Meet with Top Leadership at CDC

Dec.—HBF joined NVHR in Atlanta for a full day of meetings on December 7 with top CDC leadership that included a morning of interactive discussions with Dr. Kevin Fenton, director of the National Center for HIV/AIDS, Hepatitis, STD, and TB Prevention (NCHHSTP), and Dr. John Ward, director of the Division of Viral Hepatitis, and his department chiefs. In the afternoon, there was a special meeting with Dr. Thomas Frieden, Director of the CDC, to call attention to the more than five million Americans suffering from HBV and HCV ... and to the fact that "hepatitis is a winnable battle, too!"

HBF Executive Director and Cofounder Joan Block Receives Philadelphia Woman of Distinction Award

Nov.—The Hepatitis B Foundation, celebrating its twentieth anniversary in 2011, is proud that Ms. Block is being honored for her professional accomplishments and personal commitment to nonprofit service through her work with the foundation.

National Pediatric HBV Treatment Recommendations Published by Expert Panel

Oct.—The Hepatitis B Foundation convened an expert panel of pediatric liver specialists to establish the first national recommendations for the treatment of children living with HBV.

Major 4,000 Patient Study Validates GP73 Biomarker for HCC Detection

Oct.—Hepatitis B Foundation scientists discovered GP73, which has been validated by an NIH registration trial with four thousand individuals as an accurate serum marker for the detection of HCC and its recurrence after surgery with higher sensitivity and specificity than AFP.

Hepatitis B Education and Screening Program Offered to Subaru Employees

Oct.—The Hepatitis B Foundation and AsianWeek Foundation partnered to offer a new hepatitis B awareness program to Subaru employees in Cherry Hill, New Jersey, and more than 10 percent of the employees opted to participate in the free screening. With the support of corporate partners like Subaru, we are now getting the word out all across the nation that together we can end hepatitis B disease and liver cancer.

Ferrari Club of America Raises $22,000 for Hepatitis B Foundation

Oct.—Dr. Ron Sicilia organized the Ferrari Club of America's 2010 Charitable Auction to benefit the Hepatitis B Foundation that resulted in raising $22,000 in one hour! Nobel Laureate Dr. Baruch Blumberg and his wife, Jean, and HBF cofounders Timothy and Joan Block attended the special banquet to lend their support to the evening's festivities.

Dr. Timothy Block Selected for 2010 Life Sciences Judges' Choice Award

Dr. Timothy Block, co-founder and volunteer president of the Hepatitis B Foundation, was selected for the 2010 *Philadelphia Business Journal* Life Sciences Judges' Choice

Hepatitis B Foundation's Summer Research Internship Program Attracts Top Talent

Aug.– The Hepatitis B Foundation's summer research internships are available for college and high school students who gain "hands-on" experience in a professional lab and learn about different careers paths in medicine, research and public health.

Hep B Free Philadelphia Launched by Hepatitis B Foundation

June—Hep B Free Philadelphia is a public awareness and education campaign—based on the enormously successful San Francisco Hep B Free campaign—that is being launched by the Hepatitis B Foundation to address the growing severity of hepatitis B and liver cancer in the United States.

Hepatitis B Foundation Opens New Research Labs

June—The Hepatitis B Foundation celebrated the opening of its new research wing for entrepreneurs at the Pennsylvania Biotechnology Center, which it created to expand and accelerate its research mission.

World Hepatitis Day Rally

May—Almost five hundred people traveled across the country for the first World Hepatitis Day rally held on May 19 at the U.S. Capitol in Washington, D.C. The rally was organized by the National Viral Hepatitis Roundtable, in partnership with the Hepatitis B Foundation and other member organizations. Three members of Congress, Mike Honda, Charles Dent, and Bill Cassidy, spoke, as well as patients who all began with the line, "I am the face of hepatitis." The bottom line is that we *cannot* be silent anymore! Too many Americans are dying from chronic hepatitis, which is preventable and treatable.

Presidential Message on World Hepatitis Day

May—President Obama sent warm greetings to all those observing World Hepatitis Day and concluded by saying, "On World Hepatitis Day, we renew our support for people living with hepatitis and their loved ones, and for those who are working to improve treatment and prevention. I wish you all the best as you join together to take action against this terrible disease."

Institute of Medicine (IOM) Issues a Comprehensive Report

Jan.—The Institute of Medicine (IOM) of the National Academies of Sciences, which advises Congress on national scientific and health policies, issued a comprehensive report that confirms that more federal dollars are needed to effectively address the needs of more than 5 million Americans suffering from chronic HBV and HCV, which are the leading causes of fatal liver cancer. Liver cancer is one of the most deadly cancers and is on the rise in the United States.

Download HBF's Recommendations for Children with HBV

An expert panel of nationally recognized pediatric liver specialists convened by the HBF published the first recommendations for children chronically infected with HBV in *Pediatrics.*

Support the Viral Hepatitis and Liver Cancer Act 2009 (HR 3974)

HBF rallied to get Congressional support for this landmark bill passed for the five million Americans living with chronic hepatitis B and C.

2009

Executive Summary of First National Hepatitis Summit Now Available

A two-day forum, The Dawn of a New Era: Transforming our Domestic Response to Hepatitis B & C, was held September

10–11 in Washington, D.C., to discuss the development of a national response to chronic viral hepatitis. Drs. Anna Lok and Eugene Schiff were co-organizers of this meeting, which included presentations from medical experts, advocacy leaders, and government officials about the accomplishments and challenges in dealing with chronic hepatitis B and hepatitis C in the United States.

HBF's Expert Panel Establishes Recommendations for Children with HBV

An expert panel of nationally recognized pediatric liver specialists convened by the Hepatitis B Foundation published the first recommendations for children chronically infected with HBV that urge consistent monitoring and referral. The recommendations were posted as an early publication in the premier medical journal *Pediatrics*.

APAMSA students assist the Hepatitis B Foundation

Sept.—Sixty medical students assisted the Hepatitis B Foundation in providing free health screenings at the annual Chinatown Mid-Autumn Festival in Philadelphia.

Congressional Briefing on Viral Hepatitis

Sept.—NVHR hosted a congressional briefing on Capitol Hill as the first of a three-part series titled "Forging a National Strategy for Chronic Viral Hepatitis and Liver Cancer Prevention." Congressmen Mike Honda (CA), Charlie Dent (PA), and Edolphus Towns (NY) were special guests.

Seventeen Students "Graduate" from Summer Research Internship Program

HBF received more than one hundred applications from students at top tier colleges for nine spots in our exciting internship program, which is a ten-week experience working in our research labs. Our program has also been extended to talented high school students. Funding is provided in part by the Merck

West Point Charitable Contributions Committee and the Charles
Sigety Family Foundation.

HBF Cosponsors Successful Congressional Screening

July—A free hepatitis screening and education event was
held on Capitol Hill at which six members of Congress—Mike
Honda (D-CA), Charlie Dent (R-PA), "Joseph" Cao (R-LA),
David Wu (D-OR), Bill Cassidy (R-LA), and Donna Edwards
(D-MD)—spoke out about the need for testing and increasing awareness that chronic HBV and HCV impacts almost six
million Americans. The public event was initiated by the Chinese American Medical Society, in partnership with AAPCHO,
Hepatitis B Foundation, Caring Ambassadors Program, Hepatitis Education Project, Hepatitis B Initiative-DC, NASTAD, and
NVHR.

HBF Patient Conference Features Expert Leaders in Hepati-
tis B

Keynote speaker Dr. Mack Mitchell, chief of gastrointestinal medicine at Johns Hopkins, presented Making Sense
of HBV Treatments at the HBF's patient conference. Additional expert speakers included Dr. Michael Fried, hepatology
director at UNC Chapel Hill (NIH HBV Clinical Research Network); Dr. Kenneth Rothstein, chief of gastrointestinal medicine at Drexel University College of Medicine (HBV and Liver
Cancer); Dr. Barbara Haber, hepatologist, Children's Hospital
of Philadelphia (Children and HBV); and Dr. Timothy Block,
HBF president and professor at Drexel University College of
Medicine.

HBF Sponsors Congressional Briefing May 19 on World Hep-
atitis Day

Congressmen Mike Honda (CA-D) and Charles Dent (PA-R)
hosted the HBF's Congressional Briefing to focus national attention on the 6 million Americans who suffer from chronic HBV
and HCV, which cause fatal liver cancer. Leaders from CDC, NCI
and Weill-Cornell Medical School provided expert testimony.

HBF President Dr. Timothy Block Elected AAAS Fellow

HBF president Dr. Timothy Block joined an elite rank of scientists who are elected by the American Association for the Advancement of Science (AAAS) for discoveries that include innovative therapeutic strategies for HBV.

Hepatitis B Foundation Announces the Election of Joel D. Rosen as Chairman of the Board

The Foundation announced the election of Joel D. Rosen, Esq., a partner at High Swartz LLP in Norristown, Pennsylvania, as chairman of the board. A board member since 2003, Rosen is committed to finding a cure for hepatitis B.

2008

HBF Cosponsors Philadelphia Conference on HBV in Asian Communities

Nov.—The Hepatitis B Foundation cosponsored a conference to inform community health and outreach professionals about the severity of hepatitis B and liver cancer among Asian communities in Philadelphia.

Young Researchers Make Valuable Contributions

July—Six college students were hand-picked to work side by side with university professors and scientists involved in the latest research designed to develop better ways to detect and treat viral diseases as part of the Summer Research Internship Program sponsored by the Hepatitis B Foundation.

2008 Patient Conference Podcasts and Presentations

June—The Hepatitis B Foundation sponsored its eighth annual patient conference with workshops in Chinese and Korean, in Los Angeles. There were more than two hundred participants from across the country.

Boy Scout Troop 542 Landscaping Project at Hepatitis B Foundation

June—Upper Dublin High School student Daniel Jungkind led a group of boy scouts to complete a landscaping project in front of the Hepatitis B Foundation to fulfill the requirements to attain Eagle Scout status.

Hepatitis B Foundation Sponsors First Friday Doylestown

May—For a third consecutive year, the Hepatitis B Foundation sponsored May's First Friday Doylestown in support of National Hepatitis Awareness Month. The community event featured O'Liver, the Hepatitis B Foundation's mascot, free literature, local bands, and NBC 10 Meteorologist Michelle Grossman.

HBF Wins Award and $40,000 for Community IMPACT

March—The Hepatitis B Foundation Outreach Team was all smiles as they received the prestigious GlaxoSmithKline IMPACT award and a check for $40,000 at the company's headquarters in Philadelphia. The Hepatitis B Foundation has been selected for the GSK Annual IMPACT Award for excellence in its outreach and education programs that have helped improve access to care.

HBF Testifies Before Congress for More HBV Funding

March—HBF President Dr. Timothy Block was invited by U.S. Representative David Obey, chairman of the Committee on Appropriations, Subcommittee on Labor, Health and Human Services, to present testimony advocating for more HBV funding in the fiscal year budget 2009. Dr. Block cited the recent Nevada outbreak as a dramatic example of how the nation faces a major public health challenge that cannot be ignored. "If we don't act with urgency, more and more people will suffer from chronic hepatitis," he said.

California's Hepatitis Awareness Day

March—The California Hepatitis Alliance (CalHEP) sponsored Hepatitis Awareness Day in Sacramento, California. The Hepatitis B Foundation is a proud member of CalHEP and encouraged everyone to lend their support to this important event. The day included outstanding speakers, including Assemblywoman Fiona Ma from San Francisco and Christopher Kennedy Lawford.

GlaxoSmithKline Awards Eight Philadelphia-Area Healthcare Nonprofits with $40,000

Feb.—The Hepatitis B Foundation was selected for the GSK Annual IMPACT Award for excellence in its outreach and education programs that have helped improve access to care.

CDC Leader Dr. John Ward Spends Day at HBF

Feb.—Dr. John Ward, director of the CDC Division of Viral Hepatitis, and Mr. Dan Riedford, deputy director of DVH, spent a day at the Hepatitis B Foundation to tour the research facilities, learn about the foundation's Gateway to Care initiative, and discuss priority issues in the control and care of hepatitis B.

Philadelphia Leaders Join the Hepatitis B Foundation Board of Directors

Jan.—New board members Kenneth Blank, PhD, and R. Donald Leedy, MBA, CPA, brought many years of entrepreneurial leadership to the Hepatitis B Foundation.

Celebrate AAPCHO's Twentieth Anniversary Gala: Cultivating Traditions of Wellness

Jan.—The HBF celebrated the Association of Asian Pacific Community Health Organization's twenty years of dedication to improving the health status and access of Asian Americans, Native Hawaiians, and Pacific Islanders at their two-day conference in Washington, D.C., capped off by an elegant fundraising gala.

2007

HBF Scientists Discover Possible New Alternative to Liver Biopsy for Early Detection and Monitoring

Dec.—Scientists working at the Hepatitis B Foundation, in partnership with Drexel University, reported in the *Journal of Virology* that they may have discovered a reliable alternative to liver biopsy for the early detection of liver fibrosis and cirrhosis, which afflict more than 5 million Americans.

U.S. Representative Mike Honda Introduces Bipartisan Bill to Boost Hepatitis B Research and Prevention

Nov.—Aiming to reduce the number of Hepatitis-B victims, Representatives Mike Honda (D-CA) and Charlie Dent (R-PA) introduced a bill that they hoped would boost immunization rates against the disease and increase funding for Hepatitis B research.

Ceremony Held to Name Joshua Feldstein Pavilion

Nov.—HBF expanded its research facilities by purchasing another building named in honor of Dr. Joshua Feldstein.

Asian American Health Conference

Oct.—Chari Cohen, MPH, senior research associate, attended the Fourth Annual Asian American Health Conference at the Center for the Study of Asian American Health, NYU School of Medicine. Ms. Cohen presented results of a needs assessment of Philadelphia Chinese hepatitis B patients and their family members. The presentation focused on the specific cultural and economic barriers these patients face when seeking specialty care for their infection.

HBF Recognized in Congress

Sept.—Congressman Patrick J. Murphy of Pennsylvania recognized the Hepatitis B Foundation in a congressional statement.

He praised the foundation for its research, education, and advocacy efforts and noted that the organization is celebrating its 15fifteen-year anniversary.

Summer High School Science Enrichment Program Launched by HBF

Aug.—Talented students interested in gaining hands-on research experience were paired with a Hepatitis B Foundation scientist and mentor to learn about biotechnology and public health. The initiative was a collaborative effort with the Central Bucks School District.

Hepatitis B Foundation Helps Get CDC $1 Million More for Viral Hepatitis

Aug.—The Hepatitis B Foundation's advocacy efforts to raise hepatitis B as a national health priority resulted in the House Appropriations Committee's approval of a $1 million budget increase in 2008 for the CDC's Division of Viral Hepatitis. Congressman Mike Honda issued a public statement during the first session of the 110th Congress on July 17 to express his commitment to increased funding for hepatitis B.

Library of Congress Symposium: Combating HIV and Hepatitis B Webcast

May—The Library of Congress hosted a symposium on the challenges of developing an HIV vaccine and ensuring the eradication of Hepatitis B, in partnership with the Hepatitis B Foundation. The event was held to coincide with National Hepatitis B Awareness Week (May 7–11) and World AIDS Vaccine Day (May 18).

HBF Launches Public Health Research Department

June—The new department added a scholarly research dimension to the foundation's outreach efforts and focused on three priority areas: disease burden, liver cancer, and co-infections. Members of the team included Alison Evans, ScD, director of public health research; Gang Chen, PhD, public health scientist; and Chari Cohen, MPH, senior research analyst.

Congressional Briefing on Asian American Health (Washington, D.C.)

May—In honor of Hepatitis Awareness Month and Asian American Heritage Month, a special briefing on health issues, featuring hepatitis B and diabetes was sponsored by U.S. Congressman Mike Honda to raise awareness in the Asian American community. Expert speakers included Dr. Timothy Block.

Library of Congress Symposium: Combating HIV and Hepatitis B

May—Distinguished health and medical experts participated in a symposium addressing critical issues on the challenges of developing an HIV vaccine and ensuring the eradication of Hepatitis B. Experts from the Hepatitis B Foundation spoke about hepatitis B.

HBF President Dr. Timothy Block Receives Lifetime Achievement Award

Apr.—The Hepatitis B Foundation congratulated its cofounder and president Dr. Timothy Block for being selected to receive the prestigious Lifetime Achievement Humanitarian Award from the Central Bucks Chamber of Commerce. He was recognized at a special awards dinner for his humanitarian efforts in addressing the enormous global health problem of hepatitis B through the creation of the Hepatitis B Foundation. The Lifetime Achievement Awards, established more than thirty years ago, recognize the most influential people, businesses, and organizations in Bucks County. Dr. Block joined the ranks of past distinguished recipients such as world-renowned author James A. Michener and Pulitzer Prize–winning author Jonathan Weiner, among other accomplished individuals and organizations.

Distinguished Scholar

Feb.—The HBF announced that Dr. Baruch Blumberg had been named its first Distinguished Scholar. Blumberg planned to meet with our researchers and public health professionals to guide them in their work. "This is an exciting time for hepatitis B research and I embrace the Foundation's goal of eradicating the disease," said Dr. Blumberg.

2006

Grand Opening Celebration of the Hepatitis B Foundation's Pennsylvania Biotechnology Center

Sept.—More than 150 VIP guests and state legislators attended the grand opening of the Pennsylvania Biotechnology Center of Bucks County. The center was created by the Hepatitis B Foundation in partnership with Delaware Valley College. At the opening, a congratulatory message from Governor Edward Rendell was read by his representative. In addition, Governor Rendell announced the approval of the new biotech center as a Keystone Innovation Zone, which includes $250,000 in operational funds.

2006 International HBV Meeting

Sept.—The Twenty-first Annual International HBV Meeting was held in Vancouver, Canada. For the second year, the Hepatitis B Foundation coordinated this prestigious meeting, which is the definitive international meeting that covers all aspects of the biology of hepatitis B and hepatitis C, including biochemistry, molecular biology, traditional virology, immunology, pathogenesis, and carcinogenesis.

Congressman Supporting the HBV Bill

June—The first-ever hepatitis B legislation—the National Hepatitis B Act—was introduced in both the House (H.R. 4550) and the Senate (S. 3558). It specifically calls for screening and immunization and an increase in federal dollars to improve hepatitis B prevention and treatment.

HBF Partners with the Chinese Health Information Center

This pilot program, funded by the Aetna Foundation, is designed to educate health care and social service providers who treat Asian and Pacific Islander (API) populations. The program helps make providers more aware of the specific health needs facing the API population (especially hepatitis B), and gives information on how to treat patients in a more culturally competent manner.

INDEX

AUTHOR'S NOTE

When the solution is simple, God is answering."
- Albert Einstein

I started developing the idea for *At Einstein's Table* in late 2010 after a Hepatitis B Foundation board meeting, prompted by my interest to communicate a long list of Hepatitis B Foundation accomplishments that have taken place over the past twenty years. After receiving approval from the board and the founders, I started my work late in January 2011. Having been given full access to the Hepatitis B Foundation board and staff, newspaper clippings, *B Informed* newsletter issues, and board minutes, I was able to access information and completed more than fifty interviews. I did my best to accurately describe events and double-checked my notes and digital recordings for accuracy.

In the interest of accuracy, the table in Einstein's Table is purely symbolic. We do know that Einstein played cards in the guest house with the owner who built the larger home in front of the guest house. They must have had a table to play cards. The table would have been located in the large open room which was where the discussions were held and food was served as the foundation was forming. The title, therefore, is purely symbolic since the actual table could not be located at this time.

I decided early on that none of my interview sources, no matter how good his or her memory, could reliably recall conversations verbatim. So when I was finished writing the book, I spent time fact-checking and shared the manuscript with key people to validate facts and weed any out errors. I appreciate the response of many family and friends to provide feedback and suggest revisions. Despite these efforts, mistakes no doubt remain. I apologize in advance and would be grateful if readers would point them out.

I hope, however, that these will be overlooked by readers as they get caught up in the inspiring story of a committed group of people searching for a cure to help bring hope to the 400 million people worldwide. Thank you.

Kimberly Walton-Jungkind

2012

www.ingramcontent.com/pod-product-compliance
Lightning Source LLC
Chambersburg PA
CBHW051455170526
45166CB00001B/259